Safety is No Accident: From 'V' Bombers to Concorde

For
All Generations of our Family

Annabel   Jane   Timothy   Johanna
Emily   Rosie   Anna   Alexander (Alex)   Alexander (Sasha)
Marlow   Ivy

Each of you will have an interesting story to tell ...

Safety is No Accident:
From 'V' Bombers
to Concorde

A Flight Test Engineer's Story

John R W Smith

Ann,

In memory/appreciation of your John. He was such a pleasure and inspiration to work with. I hope you enjoy reading about some of his exploits.

John

07-08-20

AIR WORLD

First published in Great Britain in 2020 by
Pen & Sword Air World
An imprint of
Pen & Sword Books Ltd
Yorkshire – Philadelphia

ISBN 978 1 52676 944 2

Printed and bound in India by Replika Press Pvt. Ltd.

Pen & Sword Books Limited incorporates the imprints of Atlas, Archaeology,
Aviation, Discovery, Family History, Fiction, History, Maritime, Military,
Military Classics, Politics, Select, Transport, True Crime, Air World,
Frontline Publishing, Leo Cooper, Remember When, Seaforth Publishing,
The Praetorian Press, Wharncliffe Local History, Wharncliffe Transport,
Wharncliffe True Crime and White Owl.

For a complete list of Pen & Sword titles please contact

PEN & SWORD BOOKS LIMITED
47 Church Street, Barnsley, South Yorkshire, S70 2AS, England
E-mail: enquiries@pen-and-sword.co.uk
Website: www.pen-and-sword.co.uk

Or

PEN AND SWORD BOOKS
1950 Lawrence Rd, Havertown, PA 19083, USA
E-mail: Uspen-and-sword@casematepublishers.com
Website: www.penandswordbooks.com

Contents

Prologue and Acknowledgements   vi

Chapter 1   How Did I Get Into This Situation?   1

Chapter 2   Where it All Began   13

Chapter 3   A Learning Process Second to None   25

Chapter 4   Into the Real World   44

Chapter 5   No Time for Boredom   68

Chapter 6   An Unusual Measure of Success   89

Chapter 7   Restless for a New Challenge   104

Chapter 8   From Industry to Regulation: A Different Approach   118

Chapter 9   Tiger Moth to Concorde   135

Chapter 10   New Experiences One After Another   157

Chapter 11   Never a Dull Moment   183

Chapter 12   Some Difficult and Rewarding Work   212

Chapter 13   Time to Move On? Tragic Accidents   234

Appendix I: A Brief History of the Evolution of Aircraft Safety Regulation in Europe   255

Appendix II: A Summary of the Civil Aircraft Certification Processes   257

Appendix III: Author's List of Flight-Tested Aircraft   261

Glossary   263

Bibliography   273

Index   274

Prologue and Acknowledgements

Friends and family have regularly encouraged me to write about my career and experiences of aircraft flight-testing during the 1960s, 1970s and early 1980s, when aircraft technology was progressing rapidly from the post-war developments to the extensive sophistication of the jet age. The variety of aircraft types being developed, both military and civil, was astounding; it climaxed with the incredible achievement of Concorde.

This period of rapid change necessitated equally significant advances in flight test techniques and test recording capabilities. Pushing out pre-existing boundaries did not come without an element of risk. Flight testing has always involved risk; a significant number of fatal flight test accidents had occurred in the 1950s and early 1960s. It was anticipated that the advanced understanding of aerodynamics and an increased professional approach to risk management would improve the safety of flight testing. This proved to be true, but it was never going to be a risk-free activity. During my career, I lost several close colleagues and friends in fatal flight test accidents, all on civil aeroplane types.

Numerous books have been written by test pilots about their careers but few, if any, from the perspective of an aeronautical engineer working as flight test observer/engineer in partnership with the test pilot. The titles of flight test observer and flight test engineer are synonymous: 'observer' was an historic title, whereas the change to 'engineer' reflected the technical aspect of the task.

Towards the end of my flight test career in the 1980s, the role of the flight test engineer, in development and certification flight testing, was beginning to change. Advances in digital computing capability enabled real-time processing of data and, with the advent of telemetry, this information could be transmitted to specialist engineers on the ground who were able to review it and quickly pass their conclusions and advice back to the flight crew. In this new situation, if there was an unusual flight test result, it was no longer necessary for the flight test engineer and test pilot to personally make in-flight decisions on whether to continue or vary the test programme. Before the advent of telemetry, flight testing could be a lonely occupation.

As my narrative progressed, I have tried to describe the pertinent technical details of flight test activities in a reader-friendly way. Additionally, I have used the

Glossary at the end to provide some useful explanations of the terms and acronyms used. Some readers may find it helpful to browse through this Glossary before embarking on reading the book.

All the way through the book, I was confronted with a problem of the 'time-line'. In order to keep the story flowing as smoothly as possible, some events have been moved a little from their concurrent order to group them with other similar subjects. This has minimized what would otherwise have been a somewhat disjointed narrative.

In a different respect, associated with the era, the reader may be surprised to learn about situations that would now be considered, by current philosophies, to be 'politically incorrect'; for example, the way in which issues such as travel expenses and subsistence allowances were routinely handled. In addition, this period was before the days of the 'breathalyzer' or the advent of strict drink-driving laws, and lunchtime drinking was an accepted practice.

Another issue was whether or not to always use people's actual names. I have no wish to cause offence or embarrassment. However, as this is intended to be a factual description of events, I have 'bitten the bullet' and given correct names in all but a very few cases. A good number of people with whom I flew, and other close associates mentioned herein, are no longer alive. It is particularly salutary to note that none of the test pilots with whom I flew during my time at the Civil Aviation Authority are still alive; two were killed in flight test accidents (see Chapter 13). Whether alive or not, after having sought advice from others, there is no-one upon whom I have felt the need to bestow a fictitious name, except for a few whose full names I simply could not remember.

Although much of the information in this book has come from my personal memories and my flight test logbook, there were some quite significant gaps and I have been grateful to others for helping me fill these. In particular, I am indebted to Malcolm Pedel of the Civil Aviation Authority, who managed to track down many CAA Flight Test Reports and other documents that were a great help. Sadly I have not been so fortunate in locating similar information from my time at Avro/ Hawker Siddeley Aviation; in particular any detailed flight test records from the Woodford factory after it closed in 2011.

Memories are notoriously fallible. There are several instances in which I have not been able to thoroughly corroborate my recollections and I can only apologize for any errors.

There are many people who have been invaluable in providing me with help and advice. I must start with Tony Blackman. Tony was the chief test pilot at Avro (which became part of Hawker Siddeley Aviation and latterly British Aerospace) during my time there. He gave a lot of helpful advice and encouragement, as well as confirming the validity of some of my recollections. It was reading his book about his life and experiences as a test pilot that really spurred me on.[1] The other book I found inspiring was *Vulcan 607* by Roland White.[2] It described the whole story of the successful bombing of Port Stanley's runway by an Avro Vulcan in the Falklands War. This book had a style I have tried to emulate by smoothly intermixing personal stories with the events, albeit Roland White's book was written in the third person and mine is in the first.

I am particularly grateful to George Jenks, the chief archivist at the Avro Heritage Museum. On more than one occasion he diligently interrogated his photograph archives to find appropriate pictures for me to use in this book.

There were several other pictures of individual aeroplanes and specific aeroplane types relevant to the stories which I found on the image website run by Air-Britain (ABPic). This is a valuable source of such photographs as they have well over 500,000 images online, with all rights held by the individual photographers.

Other people I must mention are David Law, Andrew McClymont, Terry Newman, Keith Perrin, Jock Reed and Graham Skillen, all of whom provided valuable information to help fill the gaps in both my memory and my logbook.

Lastly, at the top of the scale of encouragement and assistance must come my wife Elizabeth. Every time she found me looking as though I had some spare time (which was very rare), she would gently 'nag' me to get on with the writing. After a year or so, when it became apparent that progress was pitifully slow, she made sure that we had some extended periods away from the distractions of home where I could concentrate on the writing. Furthermore, with her own career spanning three decades in the aviation industry and, at the time of my writing, being in the middle of an Open University English Literature Degree course, specializing in 'creative writing', she was just the person to make constructive suggestions and proofread. In some ways, it became something of a competition: would Elizabeth finish her degree course before I finished the book? I think it came close to a draw.

Notes

1. Blackman, *Test Pilot: My Extraordinary Life in Flight* (2009).
2. White, *Vulcan 607: The Epic Story of the Most Remarkable British Air Attack Since WWII* (2006).

Chapter 1

How Did I Get Into This Situation?

I t was early in 1973. We were flying in a Handley-Page Victor from the Avro flight test airfield in Woodford, Cheshire; some 15 miles south of the Greater Manchester area. We were over the Derbyshire Dales at an altitude of about 20,000ft.

'Five, four, three, two, one, go.' I selected the switch that introduced the simulated fault into the autopilot computer. The aircraft began to roll briskly to the right. I continued the count: 'One, two.'

On 'two', the Avro Chief Test Pilot, Tony Blackman, disconnected the autopilot and rapidly applied the controls to stop the roll and bring the aircraft back to wings level.

'That reached about 40 degrees of bank,' he announced, 'no problem.'

I made a note of his estimate in the instrumentation log, injected a short code of dots and dashes into the recording system to identify the test point and switched off the recorders.

'Right,' I said, 'that's the end of the straight runaway cases; next we have the oscillatory failures to do. They will probably take another forty-five to sixty minutes to complete.'

Tony's reply was immediate: 'We don't have enough fuel to do them all, but we should make a start.'

'OK,' I said, 'but give me a few minutes to re-configure the autopilot test-box for the oscillatories.'

'Go ahead,' was the reply, 'while you are doing that, we need to make a turn and head back north towards Woodford and keep away from controlled airspace.'

Although we were conducting the testing well away from civil controlled airspace, we were under the watchful eye of the military radar controllers in the area. Tony called them to report our desire to change direction.

Their reply was not encouraging. 'There is military aircraft activity currently to your north. I will route you away from the civil airspace and try to slot you in behind the military traffic as soon as it is clear.'

It took another fifteen or so minutes before we were freed to continue testing. Tony concluded that we now needed to head directly for home, but we would try to manage two or three test cases on the way back.

Handley Page Victor K2 Tanker development aircraft XL231. (*Avro Heritage Museum*)

To explain why flight testing was being undertaken on the Handley-Page Victor so many years after its original entry into service with the RAF, some history needs recounting. The Handley-Page Victor was the third of the three 'V-Bombers' conceived in the 1950s by the Ministry of Defence (MoD) to be our nuclear deterrent. The design concept was to deliver nuclear bombs into the heart of the Soviet Union. The first of the three was the Vickers Valiant. It was a relatively simple design with limited capability, but was intended to provide a stopgap before the other two – the Avro Vulcan and the Handley-Page Victor – completed their development and entered service. The Avro Vulcan is probably the best known with its characteristic large triangular (delta) wing. The Victor was a somewhat different design concept, with its bulbous nose and slender 'scimitar-shaped' swept wing. It also had a similar-shaped tailplane mounted on top of the fin (a concept well ahead of its time). It is incredible to believe that in those years, the MoD had the budget to concurrently fund three different state-of-the-art large bombers. It was not even a design competition as all three entered production. None had any defensive armaments. They relied on the fact that they could fly higher and faster than any current fighter aircraft of the time and anti-aircraft missiles were still some years away.

The Vulcan had turned out to be the most versatile and adaptable. However, military scenarios had been going through a period of significant change with

The three 'V' bombers: Handley Page Victor, Vickers Valiant and Avro Vulcan. (*Ministry of Defence*)

the advent of more advanced capable fighters, better radars and surface-to-air missiles. Bombing from high altitude was no longer an option, and the concept of descending from high altitude close to the target to fly the last segment below the radar coverage became the strategy. This needed a terrain-following capability and the Vulcan was best suited to cope with the stresses of low-level manoeuvring. Additionally, the desire to be able to return to base and land after missions in all weathers drove the MoD to require the Vulcan to have an automatic-landing capability many years before it became an achievable goal for civil aircraft (more of this later). All of this resulted in the remaining Victor B2 bombers gradually being withdrawn from service. They were returned to the Handley-Page airfield at Radlet where they were left parked outside.

The MoD realized that they had an urgent need for an in-flight refuelling tanker of greatly increased capacity to extend the range of their operational military aircraft.[1] Many of the original Victor B1 bombers had earlier been converted for use as tankers, but their performance and fuel capacity were limiting their operational capability. There was not the time or money to develop a purpose-made design, so the MoD decided that converting the Victor B2 bombers was the best solution. The B2 had the more powerful RR Conway engines replacing the Sapphire engines and the capability to carry much more fuel.

Earlier the government had been determined to rationalize the fragmented UK aircraft industry into two large manufacturing groups: Hawker Siddeley Aviation and the British Aircraft Corporation. Handley-Page was the one main aircraft company that declined to join either of the two groups. The MoD refused to award Handley-Page the Victor tanker contract until they relented. Sadly, for them, their stubbornness prevailed. No longer having a lucrative military contract and their only civil aircraft (the twin-turbo-propeller Herald) losing out in popularity to the competition of the Avro 748 and Fokker F27, Handley-Page finally went out of business in February 1970. Very soon afterwards, the Victor tanker contract was awarded to Avro and over the next year all the twenty-four airworthy Victor B2s were flown to their new home at Woodford, where they were scattered around the airfield awaiting refurbishment.

A production line was set up in the vast assembly hangar, taking over the space where the Vulcans had just completed their final updates and refurbishments. One by one the Victors were added to the beginning of the line. A thorough assessment had been made of each aircraft to determine the amount of corrosion present and the ones with the fewer problems were introduced first. It was March 1972 when the first Victor Tanker, XL231, emerged ready to fly.

The Handley-Page Victor, in common with the Vulcan, has two pilot seats and three other crew seats, side by side facing aft, to the rear of the pilots. Unlike the Vulcan, where the rear crew were at a lower level, in the Victor the larger size of the pressurized compartment allowed the three aft seats to be at the same level as the pilots. In normal operational situations, these seats were occupied by the navigator/plotter, navigator/radar and air electronics officer (AEO).

When conducting the flight testing, the flight test observer occupied the central rear seat where there was plenty of space to his left for the location of test equipment; the starboard rear-facing seat (the Nav-plotter) was not occupied. As for most intensive test flying, the policy of minimum crew was adopted. However, the port aft seat was occupied by the AEO who operated the necessary systems that were not under the direct control of the pilots and who also helped with the navigation. As with the other V-Bombers, only the pilots were provided with ejection seats. Escape for the three system operators was extremely difficult. They had to swivel their seats and, assisted by an 'explosive cushion' inflated by a CO_2 bottle to help in the event of high g-loading, leave their seat in turn for a traditional bail-out by jumping out of the crew entrance/exit doorway on the lower port side of the fuselage nose. The first action to initiate an emergency evacuation was to jettison

HP Victor rear crew stations (observer and AEO). Emergency escape door jettison handle under yellow striped cover. (*Author*)

the door of this hatch. The handle that jettisoned the door in an emergency was located on the front (forward) edge of the table immediately to the right of (when facing aft) the central crew station. In the event of a bail-out command from the captain, the drill was for this crew member to pull the handle, which was located underneath a protective cover to prevent inadvertent operation.

As I was about to become a regular crew member on the Victor flight test programme, it was necessary to practise the evacuation drill; these drills were also repeated by the whole crew at regular intervals. Before I started flying as a flight test observer in the Victor, two of the Avro test pilots, Charles Masefield and Harry Fisher, myself and a couple of other regular rear-seat crew, Bob Pogson (the AEO) and Ted Hartley (observer), went to the RAF base at Wyton, where they had a front section of the Victor fuselage set up for crew training. After a briefing from the RAF safety officer, we all took our places in our appropriate seats. As I was in the central rear seat, it was my responsibility to operate the evacuation handle.

On the command 'Bail out' from the captain, the two pilots would carry out their ejection procedures (without actually ejecting), while we three in the rear had to go through the complete evacuation process. We were in a darkened cockpit to simulate a real emergency situation. I remember, to my dismay, my hand initially fumbled with the protective cover to the evacuation handle, but it probably only took a fraction of a second to open it and pull the handle. As occupant of the central seat, I was the first to leave. I rotated my seat to face aft, simultaneously making sure that my seat harness had released; jumped out of the hatch which had already opened and landed on a mat placed on the ground below. I rolled away quickly to avoid the rapidly-following other two. Despite my problem with the handle, we achieved a total evacuation time well up with the best results achieved by regular RAF crews and the RAF safety officer was well pleased. It is interesting to note that this evacuation procedure was similar in the Vulcan, but the more cramped confines of the Vulcan cabin made it more difficult. Also, unlike the Victor, the escape hatch is directly below the fuselage and hinged at the front. The nose-wheel leg of the Vulcan is aft of the open hatch and should the crew have to bail out with the nose-leg down, they were required to grip the edge of the hatch as they exited and roll themselves to one side, hopefully so that they would avoid contact with the leg. Not something I would like to have tried!

The engineering modifications to convert the Victor B2s to K2s for in-flight refuelling were not great. As on the B1 to K1 conversion, the wing span was reduced by about 4ft to reduce the wing bending and improve the fatigue life of the wing spar. A retractable fuel hose from the rear fuselage and one from under each wing were added and the fuel piping and pumps modified to feed these. Additionally, large fuel pods were mounted under each wing to increase the fuel capacity. Although these aerodynamic changes would require some additional flight test work, it was anticipated that we would soon be able to concentrate on the primary goal of establishing that the Victor and the extended hoses were stable enough for the receiver aircraft to couple their refuelling probe into the Victor's hose.

However, things were not to be that straightforward. When the flight testing on the first Victor K2 modified for the tanker role (XL231) began, it was soon evident that, at typical in-flight refuelling altitudes of around 25,000ft, the autopilot was not capable of maintaining level flight accurately enough for the refuelling task. It took some time and effort to locate and delve into the original Handley-Page autopilot flight test reports to find the answers. It transpired that with the Victor being a high-altitude bomber spending most of its cruising time above 45,000ft at high subsonic Mach numbers (above 85 per cent of the speed of sound), considerable work had been necessary to adapt the autopilot height-lock capability (altitude-holding) for this role. The altitude instrument (altimeter) on the pilot's instrument panel obtains its information from a static air pressure port, usually on the side of the fuselage. Air pressure reduces with altitude at a fairly constant and predictable rate. It was normal practice to use this port, or an adjacent one, to provide the same altitude information to the autopilot computer. On the Victor, Handley-Page had found that, at cruise altitudes and high Mach numbers, the pressure being sensed by this port was unduly sensitive to small changes in aircraft pitch attitude (nose-up/nose-down variations). This was due to local airflow compressibility effects (a consequence of flying close to the speed of sound) in the region of the fuselage where the static ports were located. Handley-Page had tried varying the autopilot height-lock control laws (see Glossary) in an attempt to smooth out this sensitivity without adequate success. The solution adopted by Handley-Page was to find a new location for the static pressure port that was less sensitive to the Mach number compressibility effects. They had to make several flights to measure the static pressure at different positions and a suitable new location was found.

So we now had the explanation and the answer was simple: re-connect the autopilot to the original static port. Easier said than done, as this port, being redundant on the Victor bomber, had been engineered out on the production models and now had to be redesigned back in. With this eventually achieved we were back in business, but the flight testing soon revealed that the autopilot control laws would need re-optimizing for the new, lower-altitude, lower Mach number specialized role of a tanker.

An autopilot is able to 'fly' the aircraft using electrically-powered servo-motors connected directly into the mechanical circuits between the pilot's control column and the flying controls. These are the ailerons at the trailing edge of the outer wing to control roll, the elevators on the horizontal tailplane to control pitch, and the rudder on the fin to control yaw. These servo-motors must be powerful enough to apply sufficient input to control the aircraft in all required situations. Unlike later autopilot developments, they were simple systems with single-channel servos having no monitoring system to identify faulty activation of the servos; hence the consequences of a fault in the autopilot system that would cause a servo to run uncontrollably had to be investigated.

The autopilot, manufactured by Smiths Industries, was very similar to the unit in the Vulcan; Avro had significant experience of such development work. For this purpose, a control/test-box connected to the autopilot computer was located in the aircraft at the flight test observer's station. It comprised several rows of rotatable knobs, each one having numbers around its circumference. With these, one could adjust almost every parameter in the autopilot control laws. For instance, the amount of elevator control applied to correct a specific deviation such as the height from the pre-set altitude, or a change in pitch attitude due to turbulence. The sensitivity to rates of change of such parameters could also be adjusted, as could the time delay from a deviation being detected and the initiation of the control correction. Recognition and response to a rate of change is an anticipatory element in the control law. If, for example, the aircraft is pointing more nose down than intended but the pitch angle is changing in the nose-up sense, then the autopilot should decide that the error will soon reduce and little or no corrective elevator input is required. In the same scenario, if the pitch angle is changing in the nose-down sense, then more corrective elevator needs to be applied than if the detected error is not changing. The time delays (generally no more than a second) are necessary to smooth out control inputs to avoid over-correction tendencies. Anyone who has used the cruise-control on a motor car will have experienced the result of this sort of development work. If you have the cruise-control engaged and the car starts to climb a hill, you expect the accelerator to be applied to maintain a constant speed. In a good system, it will detect the rate of reduction of speed as the car begins the ascent before the speed has noticeably reduced. It will then react to this and start to increase power early enough to minimize any initial speed underrun.

In February 1973, we commenced the autopilot development testing of XL231 with the modified static pressure ports. As mentioned earlier, this involved a systematic optimization of the variable parameters in the autopilot control laws in both pitch (using elevator inputs) and roll (using ailerons). For all aeroplanes, a change of heading is accomplished by rolling the aircraft into a banked turn (generally with 30 degrees of bank for large changes of heading; smaller military 'fast jets' regularly use up to 60 degrees of bank). The rudder is not used as a primary control in normal flight, but some application of rudder is usually necessary in a turn to eliminate the tendency to sideslip and therefore rudder inputs are included in the autopilot control laws. However, in the event of an engine failure, a rapid and powerful rudder input is necessary to maintain directional control.

The primary objective of the flight test programme was to ensure that the Victor's extended refuelling hoses were stable enough for the receiving aircraft to connect with. However, there was no point in commencing any such in-flight assessments with a receiving aeroplane until the Victor autopilot was shown to be capable of maintaining a steady and stable flight path; hence there was considerable pressure to complete these autopilot tests as rapidly as possible. Much valuable time had already been lost over the static pressure ports problem.

Tony Blackman, the chief test pilot, was project pilot and I was the nominated flight test observer. We had flown together on many programmes including a lot of autopilot work on different aeroplane types and had developed a good working relationship. Tony had an extremely rapid thinking process and would often change his argument halfway through a sentence. Generally, I was in tune with his thinking and between us we would discuss suggestions for adjustments to the autopilot gearing settings which we thought would lead to the best results. After a few flights, we felt that the autopilot settings we had reached had resulted in a very capable autopilot performance.

Before going any further, we needed to be sure that the settings we had chosen did not make the autopilot so responsive that it might put the aeroplane into a potentially dangerous situation in the event of a possible autopilot system failure. The autopilots of this era were relatively simple and had virtually no failure monitoring systems. There were two worst-case scenarios to be considered. The first and most obvious case was any failure that would cause the autopilot computer to send the maximum signal to the control system servo. These cases, known as 'runaways' or 'hard-overs', were easy to replicate in flight by simply injecting a maximum signal input to the relevant control servo. The other failure scenario was less obvious but could be just as critical. When the autopilot sends a signal to the servo, the computer needs to know that the servo is following the command so that if it is not having the desired effect on the aeroplane's flight path, it can decide to either prolong or reduce the input to the servo as necessary. If this feedback is not present, the autopilot computer will rapidly conclude that more input is necessary. However, it soon detects that the flight path is changing more than intended and it rapidly reverses the direction of input to the servo in an attempt to regain proper control. Without the feedback, the same sequence happens in the opposite direction and the aeroplane is subjected to an oscillatory flight path until the pilot disconnects the autopilot. These cases are known as 'oscillatory' or 'cut-feedback' failures. It could be quite disconcerting to experience the aeroplane rapidly rolling from left to right or pitching nose up and down. Generally, if the 'hard-overs' were acceptable, the 'oscillatories' would also be found to be OK.

In carrying out the failure tests, the pilot was expected to wait for two seconds from the input of the fault before taking recovery action. The two seconds was an accepted standard 'recognition and reaction' time for the pilot to detect that something was wrong before disconnecting the autopilot and initiating the recovery. The test was deemed acceptable if during the failure and subsequent recovery the maximum bank angle reached was no more than 45 degrees for aileron runaways or plus or minus 0.5 g for the elevator cases.

Back to the story. We had successfully completed the first test programme of runaways and were heading back to Woodford with the intention of ticking off one or two oscillatory cases on the way.

Tony suggested we start with the elevator cases and when he had the Victor set up for the first test configuration, he said 'OK, let's go.'

As before, I started the count 'Five, four, three, two, one, go' and initiated the failure test.

The Victor pitched strongly nose-up, which in an instant was followed rapidly by the nose going down and back up again. The downward peaks were into negative g, albeit for just a fraction of a second. The elevator was clearly travelling from full up to full down with an alarming frequency.

'Christ! ...The tail is coming off.' I am not sure if I actually said these words but, in the fraction of a second it took to recognize the potential disaster, it was almost certainly just my thought.

I have no idea whether it was Tony disconnecting the autopilot or me deselecting the failure that came first. Before the motion had subsided, I found my right hand had already raised the cover over the emergency evacuation handle, waiting for what seemed likely to be the inevitable call from Tony to bail out. There was a short period of tense silence and I could sense Tony pausing...was there any sign of structural damage affecting the Victor's flight path? He gently moved the elevator control backwards and forwards.

'There seems to be no obvious sign of anything broken,' he said calmly. We all began to breathe again.

Tony called the Woodford control tower, briefly explained that we had encountered a problem and requested a straight-in approach. His flying was a perfect demonstration of gentleness right down to touchdown. Because we were so light in weight, there was no necessity to deploy the braking parachute from the tail; this would have been normal procedure.

After taxiing in, we thankfully exited the Victor in the normal fashion. Tony called all the relevant staff and engineers together for an immediate debrief. The first thing on the action list was to organize a careful and detailed structural inspection for any damage. An explanation had to be found for the totally unexpected, near catastrophic outcome of the autopilot failure test. There were not many likely possibilities. Could there have been a fault in the autopilot test-box or an unknown fault in the autopilot itself? Was there an error in the pre-flight weight and balance calculations? This could have meant that we were flying at a centre of gravity (c.g.) much further aft than the limit, making the aeroplane a lot more sensitive to control inputs. Everyone left the meeting, each with a determination to thoroughly follow up their actions but each hoping that they would not find the problem within their own personal area of responsibility.

I requested priority on the processing of the instrumentation recorder trace and an hour or so later, I was staring at the violent movement of the elevator and resulting pitch attitude oscillations on the recording. It was no less frightening than in real life; the whole event was over in a little more than a second. We had been very fortunate.

The structural inspections took a day or so but, amazingly, no structural damage was found; a testament to the strength of the Victor and our reflexes in minimizing the duration of the event.

No fault was found with the autopilot or the test-box. The weight and balance calculations were also rechecked and confirmed to be correct; no surprise as our weights engineers had an impeccable track record of always being accurate. However, their weight and balance document for the flight included guidance on how to keep the centre of gravity in the desired location: at or close to the aft limit. There were several separate tanks in the Victor located throughout the wings, and the recommended order in which the tanks should be used to supply fuel to the engines during the flight was specified. However, there was a limit to which this procedure would achieve the objective. When the total remaining fuel became low, there were only a few tanks that still contained usable fuel. It had not been envisaged by the weights engineer that the duration of this test flight would have been as long as it turned out to be and the fuel tank selection guidance had come to an end. After we were finally freed by the ground controller, the only available fuel remaining was in the forward wing tanks and without realizing it, as we returned towards the airfield at Woodford, the centre of gravity was moving steadily aft beyond the defined limit, towards a less stable situation. At the time of our oscillatory test, we were significantly aft of the limit, and the sensitivity to control inputs, being much greater than allowed, had caused a fairly routine test to become near catastrophic.

It was a salutary reminder that, although the primary purpose of flight testing is to ensure that the aeroplane will be safe to fly throughout its defined operational role with minimal probability of an accident, it must never be forgotten that the safety of any individual test carried out in the pursuit of achieving this goal must also be carefully considered.

How, I asked myself, was it conceivable that a test pilot who was probably the most experienced in autopilot development, and a flight test engineer, among the most experienced, should get themselves into this situation? Clearly, I thought, a touch of complacency had crept into my approach. However, secondly and more importantly, we had been in the position of not having all the information we needed to ensure that conducting the test would not be 'unsafe'. We had followed the fuel usage plan to maintain the correct centre of gravity as far as we could, but we had no direct way of knowing how far aft it was moving. There was a lesson to be learned. From then on, I would take a more direct interest in the weight and centre-of-gravity calculations when undertaking tests at the forward or aft c.g. limits. That did not mean I needed to dispute the basic weight and balance calculations of the weights engineers, but I never again went on a critical flight test without having the information necessary to determine the actual centre of gravity and ensure that it remained within the limits. This could be a fuel usage schedule on aeroplanes with multiple tanks, the movement of ballast carried in the fuselage. Water ballast tanks were frequently installed and the c.g. could easily be moved by pumping water between tanks. On smaller aeroplanes, it was sometimes possible to use myself or other crew members as a source of potentially moveable ballast.

A few days later we were back in the air and, after a few more flights, the schedule of autopilot testing was satisfactorily completed.

Note

1. Blackman, *Victor Boys: True Stories from Forty Memorable Years of the Last V Bomber* (2012), p.25.

Chapter 2

Where it All Began

I was born on 1 August 1941 at Macclesfield, Cheshire, in war-torn Britain. It's hard to be certain where my insatiable interest in aircraft began. My very first recollection of anything to do with aeroplanes was when my father attempted to make me a model aeroplane. I must have been about 5 or 6 years old and I had probably been gazing longingly at model aeroplanes in toy shop windows. I can easily envisage the likely outcome of my mother persuading my father to buy one.

In those days, back in the late 1940s, model aeroplanes had to be very light as the only means of propulsion was an elastic band driving a propeller. Therefore, they had to be constructed from lengths of very thin light balsa wood, cut to size and glued together with 'balsa cement'; the whole thing was then covered in very thin tissue paper that had been lightly damped. After leaving it to dry, the tissue gently shrank, adding rigidity and shape. The balsa wood strips were $1/_8$ inch square and other components were cut from $1/_{16}$ inch-thick sheets of balsa. They had to be laid out on a plan and pinned in place to follow the correct curve and shape of the fuselage section or wings and secured with a small dab of the cement in each joint. Once set, each section of fuselage was joined together and covered in the tissue; similarly with the wing. It was usual to hold the wing onto the fuselage with light elastic bands so that they would easily separate to minimize damage on landing.

Obviously, this process took a lot of care and precision to ensure the aeroplane was free of twists and distortions that would prevent any possibility of a successful straight flight. Within a few years, the model kits were enhanced by the addition of some components pre-cut from marked balsa wood, but still with elastic band power. It would be some time before model aircraft engines became readily available and construction methods could then become more substantial.

My father (known as Stan, although his first name was Herbert) was by far the most patient person I have ever known. Although I inherited some of this welcome characteristic, there have been many times when I wished I had his level of tolerance. He was 6 feet 4½ inches tall. I remember him being referred to as the 'gentle giant'; a recognition of his personality as well as his height. He was also a very practical person; the son of a farmer in Goole, Yorkshire. In his teens after the First World War, as well as all aspects of looking after animals and crops, he rapidly became adept at the maintenance of farm machinery. At that time, tractors and other machines were just beginning to replace horses and horse-drawn farm implements. The farm had its own blacksmith's shop which was largely the domain of my father. Among other things, he was also responsible for making and fitting horseshoes. He and his elder brother Leslie were destined to carry on the running of the farm, known as Goole Grange, after their father eventually relinquished the management. However, several unfortunate events were to influence their fate. Potatoes were one of the main products of the farm and after the infection of potato blight devastated the crop, they were in big trouble. There was also a litigation problem about which their father was badly advised. The loss of the case added to the seriousness of their financial predicament to the extent that, in the 1930s, the farm was declared bankrupt. Some of my father's uncles got together to take over the farm under a management company known as Goole Grange Estates.

A large percentage of the farm acreage was a peat moor. This was of little use for farming and furthermore, it was not the dark peat that could be used for burning as fuel. However, this type of peat, when dried and powdered, was found to make a good packing material for the transportation of delicate items (polystyrene granules being the modern equivalent). My grandfather's family and others had formed a separate company to handle this business, known as the British Peat Moss Litter Company. This company was having significant success and had acquired other land, in various areas of the country, where similar peat deposits were found.

After the demise of the farm, my father's brother Leslie went to manage one of these sites known as Creyke's Sidings, which was close to Goole. However, my father continued to work on the farm for a while, but I think that my grandfather decided there was no future for his son Stan in this and he needed to embark on another career. Radio and wireless, being the innovations of the time, seemed like a suitable choice, and so he was enrolled in a 'Wireless School'. However, from little snippets I heard from my father and others over the years, he never really had the aptitude for this and after a while it was decided, in the late 1930s, that he should follow his brother's course and was sent to manage the peat moor at Danes Moss, a site near Macclesfield in Cheshire. He had just married Marjorie who, two years later, would become my mother. It is hard to imagine how he was able to cope with such a combination of changes to his previously-known lifestyle. As a farmer's son, his management experience would be somewhat limited, and how he was able to learn and accomplish the administrative and financial side of running such a business at short notice is hard to envisage.

Danes Moss comprised 330 acres of swampy peat moor, drained by a series of parallel dykes in between which the peat was cut into turves, laid out to dry, and subsequently assembled in stacks which were rebuilt in increasing sizes as the drying process took place. A miniature rail network, built on the strips of land raised above the dyke system, was used to transport the dried peat turf to the processing mill. Each time this was done, temporary track had to be laid on the land between the dykes and joined into the permanent track via an equally temporary wooden bridge structure over the dyke. My father was entirely responsible for overseeing this construction to ensure it would support the heavy wagons full of peat. A substantial Shire horse called Bonny was the motive power for which, again, my father was solely responsible. Sometimes he would enlist the help of my younger brother Ivor and I to 'muck out' the stable.

The 'office' from which my father managed this enterprise was a wooden shed by the rail track with windows along one side. At the end opposite the door, there was a high 'Dickensian-style' desk at which he sat on a tall chair to accomplish all the necessary paperwork. This included working out the weekly pay for each of the twelve or so workers, collecting the required amount from the bank in Macclesfield every Friday morning, completing the payslips and putting the cash in individual envelopes. No electricity and no telephone, an old Imperial typewriter, a paraffin heater to keep warm in winter, and paraffin lanterns for dark mornings and afternoons: these were the limits of technology at the time.

Father's horse 'Bonny' at Danes Moss in front of rail-tracks, 1959. (*Author*)

In contrast to our father's gentle, patient nature, his life of manual work and some hereditary influences had endowed him with the biggest hands and thickest fingers I had ever seen. This was not at all helpful when trying to accomplish the delicate construction of a model aeroplane! I watched him with equal amounts of admiration and expectation as he struggled to make progress. He started with the fuselage sections and had made some attempts to join the two sides together with crosspieces without inadvertently introducing a twist into the completed structure. After one more unsuccessful attempt, on the only occasion I ever witnessed in my whole life, he lost his patience and hit it with the back of his hand. Needless to say, the fuselage was a write-off. I remember crying for a very long time with both my mother and father trying to console me and Dad trying in vain to resurrect the pieces. I am sure I was more upset to see my dad give up in such a dramatic manner than about the loss of the aeroplane as I had probably begun to realize the immense difficulty of the task. I think the remains were quietly disposed of as I never saw them again.

My brother Ivor and I spent many happy hours with our father at Danes Moss. My first vague recollection was sitting on a pannier seat behind him on his specially-large bicycle. My brother sat on another small seat mounted on the crossbar. It was a journey of about 3 miles and, at times, my brother would fall asleep. This necessitated my father rolling the bike from side to side, as he pedalled, so that he could use alternate knees to keep Ivor from falling off his small saddle.

The peat processing mill was, to me, a formidable building on two levels. On the ground floor, there was a very large three-cylinder Lister diesel engine to drive the machinery. It had to be started with compressed air from two large cylinders. Additionally, a smaller single-cylinder diesel engine, which was manually started by a hand crank, pumped air into these cylinders. It took several hours to reach the required pressure in the cylinders. The massive Lister engine flywheel had holes around its circumference. Dad inserted a huge crowbar into a convenient hole and swung on it, gradually rotating the flywheel into a position to align a mark on it against a pointer on the engine casing. The air was then released, forcing the engine to turn over at a sufficient speed to start the engine. At the other end of the ground floor, there was a covered section of the railway track along which the wagons with the dried turves of peat were rolled in and their contents manually thrown into a large hopper where they dropped into the first of the two-stage powdering machines. The powdered peat was transported by a leather belt-driven conveyor to the upper level where it was deposited into a large hopper above the baling press. This press was a substantial machine that took three people to operate efficiently. When the press was open, one man laid three laths of wood at the bottom; the sides were then closed, and the required quantity of powdered peat was released into the press. The top section pressed down to compact the bale. After releasing the top and side presses, three more laths were laid on top of the bale and three wires were looped around the laths and bale and tensioned to keep it all together. It was then kicked out of the press and manhandled out of the way to start the next one. This was no easy task as the completed bales weighed 2cwt (ten bales made 1 ton, so in metric units they each weighed about 100kg). All this machinery was driven from the Lister diesel through a very complex series of pulleys and leather belts. When the mill was fully operational, the noise was incredible.

The mill was right next to the main railway line from Manchester to London and had its own siding. The completed bales were loaded into wagons (manual work again), from where they could be transported easily to all parts of the country. The maintenance of all this machinery was entirely down to my father. Between the engine and the powdering shed, there was a large workshop with a selection of tools and a forge where all the work was done. As there was no electricity in the building due to the remoteness of the site, all the machine and hand tools were manually-powered, including a very large bench-mounted vertical drill that was hand-cranked. With no electricity, on dark winter afternoons paraffin lamps were the only source of illumination. As we grew older, Ivor and I loved to be around on Saturday mornings and in school holidays; sometimes we could be of some help to our father. This was an almost unique and invaluable experience for any budding engineer. In busy periods, my father needed to employ an assistant. Alan was a great guy and my father relied on him quite a lot. My brother and I got on with him very well. He showed us how to do many things and on one occasion he helped us to make some rather lethal 'guns' that fired ball bearings with a powerful spring. I don't think Dad would have approved if he had known. However, it was probably very fortunate that we soon ran out of ball bearings and the firing mechanisms began to wear out.

After the demise of my father's early model aeroplane-building experience, my interest in and fascination with aircraft was not diminished. As I grew older, I kept a notebook in which I noted down every aircraft that flew over. In those days, the number of different types, both civil and military, was not great; they generally flew at lower altitudes and most were readily recognizable to an experienced 'plane-spotter'. I did this for several years, but rarely could I see the actual registration marks as I did not have binoculars or very good eyesight. So strictly this was not proper 'plane-spotting'! My father occasionally took my brother and me to see the planes at Ringway (now Manchester) airport and, on such occasions, registration marks were duly recorded.

My model aeroplane-building blossomed from slow beginnings. Things were tight financially for my parents and my pocket money was limited. However, having many aunts and uncles meant that birthday and Christmas presents were a good source of extra money; providing, of course, they did not feel it would be better if I had a new pair of socks! I made little gliders from thin balsa sheet, carefully adjusting the angle of the wing (incidence), and the amount of Plasticene nose weight to achieve straight flight. This was my first trial-and-error introduction to aeroplane stability. Firstly, the centre of gravity (balance point) had to be somewhere about one-third of the way from the front of the wing (leading edge) to the back of the wing (trailing edge). Secondly, the angle of the wing to the airflow (angle of incidence) had to be a few degrees more than the angle of the tail. Any significant variation from the optimum and the glider would either try to do a loop or plummet towards the ground. Thirdly, for straight flight, the wing and tail had to be parallel when looked at from the front or back, and the fin had to be vertical and lined up straight with the fuselage.

Soon I was undertaking the construction of an elastic band-powered aeroplane very like the one my father had attempted years earlier. Although my fingers were a more appropriate size for the task, I found the same problem my father had encountered: keeping twists out of the structure. I remember persevering for a long time with some success, but several hard landings necessitated repair work, which always seemed to introduce more distortion.

Interspaced with model aeroplanes was my other enthusiasm: Meccano. This was a real 'nuts and bolts' construction kit for kids, using various sizes of metal components with holes in for the bolts to join them. There was also an amazing variety of cog-wheels and sprockets that could be used to put together quite complex machines, using an electric motor for motive power. Bulldozers, mechanical diggers and even a car with a working gearbox were some of the models I remember assembling. This tended to be a winter activity, leaving the periods of better weather for model aeroplanes.

Later, when I must have been about 12 years old, two things happened that had a significant, positive effect on my model-building success. My father set about building a wooden shed at the bottom of the garden to be used as a workshop. To my personal satisfaction, he allowed me to help greatly in its construction. Although it was quite large (about 8ft by 12ft), it was not long before I claimed a large part of it for model aeroplane construction. This meant that I did not have to repeatedly clear away the model-building from the dining table before mealtimes.

The second thing was that model aeroplane engines were becoming generally available at a reasonable cost. Reasonable, that is, to an adult. For me, I needed more than birthday and Christmas money so, with my father's encouragement, I became a paper-boy, delivering newspapers for the local newsagent. Getting up early before school was hard and the extra weight of the Sunday papers meant, on that day, I had to do the round twice. For this, I got 10 shillings (50 pence) a week – a massive supplement to my pocket money – and so, after several months, I could afford an engine and a suitable model kit.

Although balsa wood was still the primary construction material, the extra power of an engine meant that thicker sheets could now be used. Also, some parts made of thin plywood had to be incorporated for attaching the engine. The tissue paper used to cover the structure was also thicker and, instead of water, it was painted with dope (a liquid based on cellulose acetate) which when dry, shrunk slightly and resulted in a strong and quite resilient surface. The available engines were compression ignition (i.e. diesel) and of sizes from less than 1cc capacity up to about 5cc. They used a fuel that was a mixture of ether and oil; the ether being the motive power which readily exploded under compression and the oil for lubrication. Inevitably, my first such engine was of the lower size: 0.75cc. These engines were started by flicking the propeller with one's finger while adjusting the compression and fuel mixture until the right combination was achieved. The inherent danger was that the engine would often 'backfire' if the compression and mixture combination was too far wrong and the propeller would jump back and hit one's finger, inflicting a bruise. It was not so bad with a small engine, but with larger engines it could be an extremely painful experience.

The most successful aeroplane configurations for this size of engine could best be described as powered gliders. After several months of varying degrees of success, I was ready to try something new. Control-line model aeroplanes were beginning to be popular. The concept was to have direct control of the up and down motion of the model through a pair of thin stainless steel wires that entered at the wingtip. These wires were up to 50ft in length. Within the aeroplane, these wires were connected, via a mechanism, directly to the elevator control at the back of the tailplane. At the other end, the wires terminated in a control handle operated by the model-flyer, who stood in the centre of the circle as the model flew around. The model-flyer, by up and down angular movements of the handle, could thereby control the aeroplane's vertical trajectory. It was not long before such models were developed to be aerobatic, with loops, figures-of-eight, inverted flight and all manner of such combinations being possible. This type of model required more power and I needed to upgrade to a 2.5cc engine.

During the model-flying sessions at the local field, I had come across a chap by the name of Gig Eifflander. Gig, it transpired, had recently started a business in Macclesfield known as Progress Aero Works (PAW), manufacturing 2.5cc model engines to his own design. He heard that I wanted a bigger engine and he realized that, as a youngster, I did not have enough cash for a new one. He offered to sell me his prototype engine at a greatly reduced price; a kind offer I could not refuse. This engine was mechanically identical to the production versions that were beginning to be available in the shops and only differed slightly in its outward appearance. At the time, these engines were soon to gain a national reputation for power and reliability. What a lucky break; I now had the potential to move to the next level.

My brother Ivor, being just over two years younger, was also growing into a model aeroplane enthusiast, although he also tried his hand with model boats. However, there was a potential problem with boats: they could sink and never be seen again! The worst that could happen to an aeroplane was a severe crash, but generally there was enough left to attempt a rebuild. So the two of us would spend hours together in the shed, building our various models. Mother would often come down from the house to bring us a drink or tell us it was a mealtime. In winter we kept the shed door closed and when Mother opened the door, she would refuse to come in. The combination of the acetone in the balsa cement, the dope for the covering and the ether in the engine fuel created a powerful concoction. With hindsight, it would have been a glue-sniffers' paradise. Later, I often wondered how much this atmosphere contributed to our enthusiasm for model-building!

Over the next few years our models became larger and more capable. I acquired some larger engines of 5 and 6cc capacity. These were 'glow-plug' engines rather than the earlier compression ignition types and were somewhat like car diesel engines in principle. To get them started, the element in the glow-plug fitted in the cylinder head was connected to a 2-volt rechargeable lead–acid battery, similar to a small car battery. To start these engines, the same technique of flicking the propeller was required but, once started, the residual heat of the combustion would continue to ignite the fuel and the leads to the glow-plug were then removed. Instead of the ether/oil mixture, these engines used a mixture of paraffin, methanol and oil with the addition of a small quantity of something called amyl nitrite, which seemed to be an important component. This appeared to be quite a lethal liquid and, although the ready-mixed fuel available in the model shops contained this, we wanted to save money by mixing our own fuel. The paraffin, methanol and oil were readily obtainable so Ivor and I, with me barely in my teens, decided to venture into a local chemist's shop to see if we could buy some amyl nitrite. In those days, chemists' shops were as described: shops that sold chemicals, although this did include medicines and some cosmetics. We explained why we needed it, and to our amazement the chemist produced a large bottle in one of those papier-mâché protective boxes and happily sold it to us for not a lot of money. It lasted us throughout our model-making time and I remember finding the remnants in the same bottle very many years later when we had to clear out the house and shed after our parents had died. It still had the same distinctive smell which brought back happy memories. I later looked it up and found it described as 'a highly volatile flammable liquid' which 'can cause serious harmful effects if inhaled'. How unlikely would it be now, even for an adult, to buy this over the counter?

With the larger engines, my models became bigger, more capable and more substantial. For me, the main focus was control-line aerobatic aeroplanes. These 5 and 6cc engines were powerful enough to keep the control wires to the model from going slack, even in moderate winds. To increase manoeuvrability, I fitted flaps at the trailing edge of the wings, which moved down when the elevator

The author with control-line aerobatic model aeroplanes, 1960. (*Author*)

moved up and vice versa. Apart from the obvious advantage of improving the manoeuvrability, by increasing the wing lift as the elevator went up and reducing the lift as the elevator went down, this moved the centre of lift aft which had the secondary effect of reducing the stability and so increased the responsiveness during the manoeuvre.

The next natural progression was to test my ability in model competitions. Ivor and I had both become members of the Macclesfield Model Aircraft Club. Gig Eifflander was the driving force and chairman. At one of the meetings, I tentatively suggested that the club should organize some competitive events for the local members. This was greeted with encouragement but – surprise, surprise – I was the only nomination for the post of competition secretary. With more enthusiasm than common sense, I set to work on planning a series of competitions for different categories of models. Many of the other members were working people with families and, compared with Ivor, me and some of the other younger members, they had limited time for constructing more than one or two types of model.

The competition programme I planned included three variations of the control-line theme: firstly aerobatic; secondly speed, where one tried to insert a powerful engine into a relatively small and aerodynamic model, hoping it would stay airborne long enough to complete three circles that were timed; and thirdly combat, where two flyers in the same circle would do their best to cut off pieces of a streamer towed behind each other's model. Then there was a selection of free-flight competitions: firstly, tow-line gliders where the model was towed into the air with a fixed-length tow-line and its duration timed from the release of the line; and secondly 'chuck-gliders', which generally had wings, fuselage and tail made from sheet balsa, and were launched by hand in a curved flight trajectory to reach the highest possible altitude for maximum gliding duration. Thirdly, there was another variation of gliding duration: this required an engine to propel the model to altitude using a set amount of fuel where, if it all worked out according to plan, the model would be in an optimum attitude for gliding. Ivor was the master of this, at least among our members. The engines for this class were limited to 1.5cc; otherwise the model could disappear upwards, almost out of sight.

Nevertheless, the members tried their best to be involved in as many of the competitions as possible. Gig was the arch competitor. With a total of only two models plus a rapidly-constructed chuck-glider, he entered all the competitions. His control-line aerobatic model was quite small, and he also successfully used it for the combat competition and, with a specially-tuned engine, it was fast enough to beat all but Ivor's specialized speed model. He had a powered glider from

The author's brother Ivor launching powered glider, 1960. (*Author*)

which he removed the engine and replaced it with a lead weight for the tow-line competition. He did not win any individual competition, but he was second or third in all the events and his overall performance was good enough to beat everyone else; a salutary lesson of experience over the enthusiasm of youth.

The competitive bug had bitten several members and a hard core of us set about entering several national competitions. My preferences were control-line aerobatics and combat, while Ivor tested his skill with the powered glider; he also tried the control-line combat. By this time I had learned to drive, passed my test (at the second attempt), and Father was kind and trusting enough to lend me his car to take us and our models to various events. They were invariably located on military airfields with sites set aside for pitching tents. We had obtained an old army bell tent that would accommodate four or five of us, but in order to have room for the other necessary camping equipment and our models, we acquired a small trailer. Father kindly fitted a tow-bar to his Morris Cowley and off we went. Needless to say, at this age I was not an experienced driver and towing a trailer was an additional learning curve to be assimilated. We went to several such events and enjoyed ourselves immensely, but never really got close to winning anything before it was time for me to move on to apprenticeship, university and motor cars.

Chapter 3

A Learning Process Second to None

While Dad got on with being the breadwinner in his capable but easy-going style, Mother was the driving force and organizer. She knew very well that the most important thing for her children had to be a good education. Mother had realized that only a few children from the local primary school made it into the grammar school. She did have some money put aside inherited from her parents, and at the age of about 7 or 8, I found myself having private tuition on Saturday mornings and in school holidays from teachers at the junior section of the Macclesfield Grammar School. Mother's objective was for me to take and pass the entrance examination to this junior school. I don't remember taking the exam, but I was clearly successful. The two years of education provided by this school was primarily dedicated to passing the '11-plus' examination in order to get into the grammar school. For me this could have been potentially more difficult than average because, with my birthday being on 1 August, I was nearly always the youngest boy in the class. However, the 11-plus exam consisted of three sections: Maths, English and an intelligence test, all in a multiple-choice format. This clearly suited my logical way of thinking and I achieved a very good pass.

So in September 1952, having just reached my 11th birthday, I entered the King Edward Grammar School, Macclesfield in the top class. My mother's efforts had resulted in the best possible outcome. Thereafter it was up to me. My brother followed the same path three years later.

In an attempt to give us all a balanced education, the school curriculum was designed so that we would arrive at the Ordinary or O-level (now GCSE) exams five years later with some variety of subjects. Mathematics, English Language and a foreign language were not open to choice. In the first year (designated the third form in those days), I do remember having lessons in History, Geography, Science, Latin, Music and Sport. I don't think the French lessons started until the following year. Maths and Science were a pleasure. Geography was of some interest when dealing with maps and the formation of geological features, but when it came to remembering the names of cities, towns, rivers and mountains, I was not enthusiastic. Similarly, History started with the most fascinating pre-historic periods, but soon degenerated into memorizing the inevitable lists of dates and historical characters. As far as I was concerned, Latin was indeed a 'dead language'; I just could not see any point in having to learn endless words that were of no use at all in the real world! When it came to Music, I thought I might enjoy being able to play a musical instrument but, no matter how hard I tried, I had no aptitude for translating music into notes on any kind of instrument. The only instrument with which I was entrusted was the percussion triangle. I did have some interest in being part of the choir, but it did not take the music teacher very long to determine that I had no singing voice, so that was the end of that.

I was more enthusiastic about sport. My father was a very keen cricketer in his younger days and he played with us on any piece of flat ground we could find. There was a small field, sometimes occupied by cows, close to our house. There was a fairly smooth area and I remember Dad wheeling his lawn-mower round to the field to mow a pitch. So cricket was quite fun and messing about in the gym was OK. The main winter field sport was rugby, but at that time I was just about the smallest boy in the class, which was something of a handicap despite my initial enthusiasm.

So my curriculum horizons were shrinking fast. At the end of the first year, although I did well in Maths and Science, I only just managed to scrape high enough marks in the other subjects to stay in the top group for the second year. This knife-edge path to success turned out to be the same at the end of the next two years.

In the fourth year (then designated the lower fifth form), things were much easier. Firstly, there was the option of splitting Science into three separate subjects: Physics, Chemistry and Biology. For me, this needed no second thought. However, together with the compulsory subjects of Maths, English Language and French, it still left me having to choose two further subjects to make up the necessary eight for the O-level exams at the end of the fifth year. A 'craft' subject was one option and I eagerly chose Metalwork. Because of the lesson scheduling, I was left with a choice between Geography and History, neither of which inspired any affinity or enthusiasm. Mother recognized the benefit that either of these subjects would have to give a better balance to my overall choice, and strongly encouraged me to take Geography. However, notwithstanding Mother's pressure, somehow I bucked up a lot of courage and went to see the senior master for the lower fifth year. Surprisingly, he was quite sympathetic as he knew from my record that I would struggle with History or Geography, but there were no other subjects that would fit into the lesson timetable. After giving it a bit of thought, he hit on a possible solution. The school had introduced a new subject into the O-level curriculum:

Mechanics. This was a sort of Applied Maths and was offered as a subject for some of the fifth forms where the three separate science options were not available. There were not many pupils who had chosen this subject option – barely enough to justify its introduction – and an extra candidate would be welcome. Fortuitously, some of the weekly scheduled Mechanics lessons did coincide with the History/Geography slots. The senior master had to discuss this with the headmaster and other relevant masters but, to my delight, it was agreed in principle, provided I understood that I would have to study, unsupervised, the curriculum covered in the lessons I had to miss. I readily agreed, and I found it such an interesting subject that it was not too difficult to keep up with the other pupils.

It is worth mentioning, in passing, the attitude to safety in the Metalwork course. In the workshop there were several machine tools as well as an open-fired forge. Apart from having to wear a light buttoned coat to protect our school uniform and gloves for holding hot metal, no eye protection was provided nor was there much in the way of protective guards on the machines. The teacher explained the dangers and left us to it with a bit of general supervision. Everyone took their own responsibility for their personal safety and I don't remember anyone suffering anything worse than minor burns or light scratches.

At the end of the fifth year, we took the O-level exams; eight in my case. I achieved good marks in all subjects except for English and French, both of which I failed. This did not prevent me from being eligible to enter the sixth form and start the Advanced or A-level course, but it was made very clear to me that unless I passed these subjects, I would not be able to go to a university. There were opportunities to re-take failed exams in the middle and end of each sixth form year. At that time, all universities required English and a foreign language as entry criteria. Additionally, all the qualifying O-level subjects had to be on no more than two separate exam certificates. This meant that I had to pass both English and French at the same time. At the first re-take, I passed English but failed French. At the second attempt, I failed English but passed French and it was at the third attempt in the middle of my final school year that, thankfully, I managed to pass them both.

There were some other O-level subjects that were available for study during the sixth-form years, one of which was Technical Drawing, which I took up with enthusiasm. Because of the 'two certificate' rule, each time I took French and English, I also took Technical Drawing and ultimately passed it three times!

I had consistently done particularly well in Maths and Physics, so these were the subjects I chose for the A-level exams. At that time, there was an option for a second Maths subject called Further Maths. This was an interesting combination of Applied Maths and more advanced mathematical concepts not covered by the normal Maths curriculum. Out of the whole sixth form, there were only eight of us studying this subject and we enjoyed being taught by the very best Maths teacher in the school, in a friendly and almost individual manner.

During my last two school years in the sixth form, I had to do some thinking about a career. It had to be something to do with aeroplanes. I was not going to be a pilot as my eyesight was not good enough; there did not seem to me any other attractive prospects in the airlines or the RAF. The only option in my mind was to join one of the main aircraft manufacturers. I liked the idea of design or research work, but with hindsight, I really did not know much about the whole spectrum of activity that went into designing, manufacturing, testing and certificating an aircraft. The obvious choice of a manufacturer was the Avro company as it was the nearest to home. They had a large design and manufacturing plant at Chadderton, just to the east of Manchester, and a final-assembly and flight-test site at Woodford in Cheshire, just south of Manchester. I was persuaded by the school's career officer not to put all my eggs in one basket and consider an alternative choice. After a bit of thought, I settled on the Blackburn Aircraft Company, which was based in Brough, East Yorkshire. At that time, they were beginning to develop the Buccaneer, a twin-jet, ground-attack aircraft. Avro was easily my first choice, not just because of its proximity to home, but because they manufactured a bigger variety of aircraft types including the famous Vulcan bomber, and the new civil Avro 748, with its twin-turbo-propeller engines, was in the early design stage.

With some further help from the school's careers officer, interviews were duly arranged. The Avro interview would take place at Chadderton and the Blackburn interview at Brough. Father arranged to take time off from work to drive me to each of these. I don't remember much about the actual interviews except that I was treated very kindly and was able to answer most of the questions with some enthusiasm. In particular, at Brough, when the personnel officer realized that my father had driven the not inconsiderable distance that morning, he invited him to sit in on the interview. This was good because afterwards Father could give me some very good advice, such as making more eye contact with the interviewer.

In the late 1950s there was so much activity in the UK aircraft industry, one would have to have made a really bad showing at the interview not to be offered a job. At the time, the favoured route to qualification was via the 'sandwich' apprenticeship. This scheme started with a one-year apprenticeship at the company, followed by a three-year university course and then a final one-year apprenticeship back at the company. As an apprentice, you were also expected to spend time during the long summer university vacations continuing the apprenticeship at the company. It was therefore not necessary to initiate an application for a university place until after starting with the company; one thing less to worry about in the last exam year at school.

Both companies offered me an apprenticeship on this basis and I accepted the Avro offer. I had particularly enjoyed the school sixth form. I was in my element with the purity of mathematics and physics. At this level things were either right or wrong. There were no theories that could not be proved; no methodologies that could not be deduced. I loved it. I was rewarded with two 'A' grades and one 'B' grade. I also had my O-level passes on two certificates, so nothing was in the way of a place at university. Interestingly, my brother, three years younger, was even worse than me at French. He re-took the exam and failed it four times during his sixth-form years and it was looking unlikely that he would be accepted by any university. In desperation, at the start of his pre-university apprenticeship, he had enrolled for private French lessons. However, with a bit of cunning and lateral thinking, he had discovered that one university, Imperial College London, did not insist on its applicants having an O-level language pass. They had an Electrical Engineering course, which was his preference, so this saved him the stress of more learning and repeated exams.

I left school in June 1959, having yet to reach my 18th birthday. The Avro apprentice training school was located at the Woodford factory site, where I started in September 1959. There were some lectures on methods and processes that were used in the aircraft manufacturing industry, and the different departments I could eventually join after graduation. However, many of these first few weeks were spent in the apprentice workshop learning how to work metals with hand tools and machines. For me, with the benefit of Father's tuition and opportunities, this was easy stuff, although I did learn a lot about machine processes. We had the opportunity to demonstrate our new-found skills by making some useful things. I still have, and still use, the hacksaw I made at that time.

Since leaving school I had given some thought to my choice of university course. Because of my love of mathematics, I seriously considered a Maths degree rather than an engineering degree. Fortunately, I reminded myself of where my true interests lay, and I chose the Aeronautical Engineering course. The two universities with the best reputations for Aeronautical Engineering were Bristol and Southampton. It appeared to me that Bristol offered the best curriculum, so I sent in my application.

Soon after starting at the Woodford Apprentice School, I went off to Bristol University for the entrance interview for the Department of Engineering; specifically, the Aeronautical faculty. I did not fully appreciate that these interviews were far from a formality, and to be totally honest, my interview performance left a great deal to be desired. With my many aircraft interests and the single-mindedness that I wanted to make my career with aeroplanes, I had a potentially strong case with which to impress. However, I realized afterwards that I had not made the best of this and had presented myself as rather casual and facetious.

A few days later, the apprentice training supervisor called me into his office and told me that I had not done at all well in the interview. Unbeknown to me, he was a close friend of one of the professors at the interview. From his experience with me, he seemed to believe I had the necessary potential and he said he would put in a supportive case on my behalf to the professor. To my immense relief this intervention succeeded, and I was accepted onto the university degree course. This was a monumental stroke of fate without which my life would have taken a different and, no doubt, less successful turn. I often think of the gratitude that I owe the apprentice supervisor and greatly regret that I never took the opportunity to express this.

After the initial basic training, the apprentices were required to gain first-hand experience in all aspects of the manufacturing processes. This required a transfer to the factory at Chadderton. It was too great a distance from home to contemplate the journey on a daily basis. The Apprentice Training Department had a pool of lodgings where apprentices could be accommodated during the week and I was allocated a temporary 'home'. Apart from occasional visits to relatives, this was my first experience of being away from my family on my own. The landlady and her husband could not have been more kind and understanding; they did their best to make me feel welcome. He worked at the Avro factory and was able to take me to and from work by car every day. Televisions were a rare luxury in those days, but they had one and I well remember the early episodes of *Coronation Street*.

The apprentices spent two weeks or so in each department. It began with cutting out aircraft parts from duralumin sheet (an alloy of aluminium with small quantities of copper and magnesium). It was mostly done by machines, but quite often required a degree of 'finishing' with hand tools. This was a very good use of apprentice man-hours.

Almost all processes were done on a piece-work basis; that is, operators were paid for the number of pieces of work or processes that they had finished. A 'rate-fixer' assessed the number of man-hours that it should take to do each element of a process; he then calculated the equivalent pay. The operator's weekly pay was determined by the number of elements accomplished in a week. It was usually possible to undercut the allotted times; then the operator would receive the difference as a bonus. This was not only an incentive to work as quickly as possible, but also for the operator to train his apprentices to help efficiently with the work. The apprentices' man-hours were paid from a separate budget; any useful work that they accomplished went directly into the operator's bonus pay. In a few of the sections, the local senior operator would recognize this contribution and pass on some of his bonus pay to the apprentices.

After several months of that 1959/60 icy cold winter, we had passed through many sections and departments. These covered sheet-metal pressing, metal casting, machining, riveting together small components, plastic fabrications, electrical cabling, protective coatings and paint-spraying. For a 'do-it-yourself' enthusiast, the waste bins scattered around each department were an 'Aladdin's cave' of useful and interesting items; it was forbidden to use any scrapped material on an aircraft. I found some lengths of thick electrical main generator cable from the Vulcan which I still use today as jump leads for my car battery. Some off-cuts of clear plastic sheet I carefully fashioned into a set of long French curves: tools used by draughtsmen for drawing smooth curves of variable radius. Again, I still have these today.

During the final weeks before starting university, the apprentices returned to Woodford to gain some direct experience of final assembly processes. This was, for me, a very memorable time and an almost unique situation in UK aviation history; three totally different production lines, side by side. The Avro Vulcan was the main aeroplane occupying the production line facility. There was also some refurbishment work being undertaken on the Avro Shackleton: a maritime reconnaissance aircraft developed from the wartime Lancaster bomber. Additionally, the Avro 748 assembly line had just started to be developed.

As an apprentice, I helped to fit together the large components, install the enormous wiring looms, and check the operation of undercarriage and flying controls. I felt immensely proud to be part of this impressive manufacturing process. On a lighter note, I have one everlasting memory of lunchtime cricket games held inside the Vulcan's bomb bay; clearly this necessitated the use of a very soft ball!

During this first year of my apprenticeship, every Friday during term time we attended lectures at Salford Technical College to keep our brains active in mathematics, science and engineering subjects. It was also an opportunity to socialize with other undergraduate apprentices who were on similar 'sandwich' courses. Our lunch break was quite long and we spent it in the local pub. This was all a new experience for me; to be honest, I took to it like a duck to water. At home, an infrequent shandy on special occasions had been my only alcoholic exposure. Looking back, I took the sensible route by starting my drinking career with half-pints of the local mild beer: a popular northern draft brew that was not too bitter and less alcoholic than normal bitter beers. After a few weeks, I progressed to pints of mixed mild and bitter; a choice I stayed with in later years wherever it was available but, sad to say, mild was rarely brewed in the south of England.

In early October 1960, aged 19, I started the Aeronautical Engineering Degree Course at the University of Bristol.

I had been slow in making my accommodation arrangements. I was too late to get a place in the halls of residence; a first choice for most students because everything was well-organized and there was a good communal spirit. Once these places had been filled, the university's administration had a comprehensive list of private residential homes that would accommodate students. I had not been left with much choice, but as I had no personal knowledge of any of the available residences, I accepted their recommendation.

I was allocated a place in a large house in the northern suburbs of the city. A young couple had purchased it with the express intension of maximizing an income from student lettings. The house was mortgaged, and all the furniture had been bought on hire purchase which they hoped to pay off in the first year. The husband had a full-time job, but also helped with breakfasts, suppers and weekend lunches. From memory, there were between twelve and fourteen students housed there. I shared a room on the ground floor with two other students who were both studying engineering; one Mechanical, the other Electrical Engineering. There were also four other engineering students in the house; one Aeronautical, two Civil and one Mechanical. The rest were a mixture of Medical, Chemistry and Philosophy students. Apart from the student bedrooms there was a dining room, a television lounge and a small kitchen. The owners lived in a small room next to the kitchen, only just big enough for a double bed. The engineers tended to socialize together. In particular, the other Aeronautical Engineering student, Geoff Tomlinson, became a firm and long-lasting friend. As the social highlight of the week, we all crowded together in the lounge to watch *That Was the Week That Was*: the first satirical television programme fronted by a very young David Frost, on a black and white screen of course!

Being students, we could not exist without an occasional party with a lot of music, beer, dancing and falling over. The owners of the house generally left us to it. The major casualties of these parties were the beds. They were cheap and had flimsy metal frames that could not support the weight of many people sitting on them. The legs buckled, and although we managed to bend them back into shape, after two or three such occasions fatigue fracture was unavoidable. However, if the

The student house residents, 1961: author at back right, Geoff Tomlinson at front left. (*Author*)

damaged bed was placed carefully against a wall it would retain sufficient stability to perform its function as a bed. The problem was that not all the beds in the house were located against a wall. If any of the engineers had a terminally-damaged bed with no wall to support it, we would reconnoitre the other rooms for beds next to walls and when the occupants were out, secretly make an exchange. We got away with this until partway through our second year when there were no free-standing beds left intact and the landlord had to replace several beds. He was not happy to have his profit reduced, so some of us were not invited back for the third year. As usual, the ringleaders escaped this 'sentence'.

The university Engineering building was a short walk from the nearest bus stop; at the other end, there was also a ten-minute walk from our 'home'. After a few days, we worked out that we could walk the whole way in thirty minutes and save money. We did this unless it was raining hard. Later some students acquired old cars to use now and then. However, in cold wet weather they tended to be unreliable and push-starts were frequent.

My first day at the Engineering faculty began with all the students assembled in the main lecture theatre for an introductory talk. The four Engineering faculties

University of Bristol low-speed wind-tunnel, 1962. (*Author*)

– Mechanical, Civil, Electrical and Aeronautical – had about thirty students in each; an approximate total of 120. Lectures for subjects such as structures, mathematical techniques and fluid dynamics would be common to more than one faculty, particularly in the first year. After the explanations and formalities, we were introduced to most of the lecturers and then taken on a tour of the facilities.

The nucleus of the Aeronautical Engineering Department was a very extensive laboratory; the dominant feature was a low-speed wind-tunnel that we would use a great deal, particularly during our third-year projects. There was also a structural test set-up; it used an actual wing from a de Havilland Venom jet fighter. It could be loaded with hydraulic jacks at various points that enabled structural stresses to be measured with strain gauges: a simplified version of set-ups used for structural testing by the manufacturing companies. This was a source of one of my biggest academic disappointments. In the third year, we worked in pairs on several projects. I had teamed up naturally with Geoff Tomlinson. With all the knowledge we had gained about stability and control equations and structural analysis, each pair was asked to plan and execute a structural test programme on this rig to examine how the wing would twist under different aerodynamic loads; then match this with the stability characteristics. I was so confident in my understanding of the issues involved and my mathematical analysis capability, that when we finally got to the

University of Bristol aircraft structure test rig, 1962. (*Author*)

end of the calculations I could not believe that the answers did not correlate as well as they should have. Although the lecturer assured us that we had done a good job and rarely did anyone achieve a better result, it left a doubt in my mind. Geoff and I went over our test results again; we decided that, with hindsight, perhaps we should have made the test programme more extensive, but each team was only given a limited time for this.

Apart from the academic work and socializing, the university had a list of clubs and activities in which the students could join. I had studied this list to see if there was anything that particularly interested me. I thought sailing might be a good idea, so I duly put my name down. I don't know what I had expected but, being late autumn, all the actual sailing had finished for the year and the boats were out of the water for their annual refurbishment. The few other 'new boys' and I were put to work sanding down the bottoms of the wooden hulls. After two or three sessions of this with no sign of any tuition in the fundamentals of sailing I lost interest, along with several of the others, and withdrew from the 'club'. I was of the impression that the second- and third-year students did not wish to encourage too many new members so that they themselves would have more access to the sailing opportunities on the limited number of boats.

Later, in the second year, I became an enthusiastic roller-skating hockey player; a highly energetic and dangerous sport carried out in a gymnasium. It was almost impossible to complete a game without some minor injury, but the worst damage I did was to crack a bone in my wrist.

The engineers generally became a well-integrated social group. We tended to congregate in the Student Union bar and attend the regular music and dance events there. At the time, traditional jazz – known as trad jazz – was the popular music; three or four times a term, Acker Bilk or Kenny Ball and his Jazzmen would entertain us. Then there was 'Rag Week' each year. Although the main objective was to raise money for charity, it was also a good excuse for a good time. In both of these activities the engineers more than pulled their weight. We spent many evenings going from house to house in the Bristol suburbs collecting substantial charity donations. This was usually followed by a visit to a local pub for a quick drink before closing time. In those days, licensing hours terminated at ten o'clock on weekdays and 10.30 at weekends, which was probably a blessing.

During the summer vacation after the first year, I began to think about buying a car. At 18, I had passed my driving test at the second attempt and although I had continued to enjoy the extensive use of Father's car, I began to feel I needed the flexibility of a car of my own. At 17, still at school, I had been keen to buy a small motorbike. Father would not have stood in my way but, with his usual wisdom, he made me an offer: 'If you don't spend your money now on a motorbike, however much you save up when you are able to buy a car, I will double.' I gave it some consideration, but concluded it was an offer not to be refused.

So now was the time to cash in on that promise. With my earlier model aeroplane activities and more recent university living expenses, I had not accumulated a big sum. However, I had a desire for something a little more exciting than the Austin Sevens and Morris Eights owned by some of my student friends. I looked around the local second-hand dealers and found what I thought was an ideal choice: a Sunbeam-Talbot Ten. It was just as old as the Austins and Morrises of my friends, but it had a little more appeal and I reckoned I could just afford it.

It turned out to be the beginning of my car maintenance learning curve. First, there were the brakes. They were mechanically operated and very temperamental. If not carefully adjusted on a weekly basis, their application would cause at least one of the wheels to lock up while the others were still turning; a recipe for a serious skid, particularly when wet. Then there was the tendency for the starter motor engagement not to release from the flywheel after attempting to start the engine. Pushing the car to rock it backwards and forwards with first gear engaged usually succeeded in releasing it. Carburettor adjustment, damp in the distributor and the usual battery running down problems were never far away. Nevertheless, I drove it to Bristol and back for each term of the second year.

The author's Sunbeam-Talbot 10 with friends and Rodney at the wheel. (*Author*)

The second year was the best for socializing and other non-studying activities; one of these was the Rag Week. Of the many events organized by the students, the 'twenty-four-hour pedal car race' was the one taken most seriously by the engineers. The only rules were that they had to be powered by only one person at a time and the wheels had to be propelled by a fore-and-aft movement of the pedals. Pedals connected to a rotating wheel such as those on a bicycle were not allowed. The Engineering Department had the best facilities for constructing purpose-built machines. This did give us an advantage, but other faculty's teams were allowed to take advantage of these facilities in order to encourage as many entrants as possible. We all had a limited budget, but it was sufficient to purchase a standard children's pedal car and make adaptations. It was during my second year that I became seriously involved with the Engineering Department entry (the academic work-load in the third year would be too demanding for such things). In fact, we had two entries in my second year. One was a purpose-designed contraption with a driver in a laid-back position that enabled him to put maximum effort into the pedals. It was decided that we needed a back-up in case our complex purpose-built car failed to survive the ordeal. For our second entry, we obtained a child's very simple, standard pedal car. We wanted to make minimal modifications to this car, but it had to have some work done in order to fit the drivers in. We selected the smallest drivers we had. The problem was, with the angle at which the driver was seated, he had great difficulty applying sufficient force to the pedals. A greater problem was the need to move the pedals back and forth at a high enough frequency to propel the vehicle at a reasonable speed. It was extremely difficult. However, we did have one driver, Rodney Hadwen, also an Avro apprentice and close friend, who mastered this high-frequency co-ordination of legs and feet with amazing effect. He ended up completing the lion's share of the driving of this pedal car over the twenty-four hours of the race.

The course was a triangular layout of pathways in front of Bristol City's council building. There were about fifteen entries of various designs. From the start our number one machine managed to keep ahead of the pack, but after several hours with regular driver changes, many of the purpose-built cars began to suffer failures, ours included. Inevitably, the main problems were wheels, wheel bearings and push-rod mechanisms to the driving wheels. Over the twenty-four hours we spent a lot of time repairing the cars to keep them running. Less than half the cars were still going at the end. The finish turned out to be a real tortoise-and-hare scenario. Our number two car kept going almost faultlessly, propelled by Rodney's high-frequency legs. In the last hour, while two or three of the remaining faster, specialized cars (including ours) were undergoing major repairs to keep them in

Rodney driving in the 'Rag Week' pedal-car race, 1962. (*Author*)

the race, Rodney had taken the lead. We got our number one car back into the race and it just managed to beat Rodney into second place a few laps before the end. Apart from during the night, this event attracted many local spectators. In between repair work, we were able to make good use of our collection buckets for the benefit of several charities.

For our final year, 1962/63, Geoff and I had to find alternative accommodation, having been 'asked to leave' the lodgings that had been our home for two years. Geoff took the initiative in locating a flat for us to rent. He found us a place on the second floor of one of the large terraced houses quite close to the university. Many such houses had been divided into flats for students. Apart from one unexpected event, this proved to be a successful residence for us.

Back home in the last summer vacation, I had my 21st birthday. I had been aware that one of my uncles had left a little money in his will to each of his nephews and nieces when they reached the age of 21. My mother had several brothers and sisters and so there were quite a lot of nephews and nieces on the list and I was not expecting very much. It did turn out to be quite a generous sum and I really did need a better car. I saw the wisdom in putting some of the money into savings, so I was back touring the second-hand car dealers for a bargain. Understandably, some were not very interested in taking my Sunbeam-Talbot in part-exchange,

The author, Rodney and friends with author's Singer Gazelle, Bristol, 1963. (*Author*)

but I found a dealer who would and they had a Singer Gazelle convertible at a reasonable price. I was very pleased to see the back of the Sunbeam-Talbot. I drove my new car with pride down to Bristol at the beginning of the third year and enjoyed this car for many years.

Shortly after our arrival at the beginning of January, for the penultimate term the weather deteriorated rapidly. Snow fell and the temperature dropped well below freezing, where it stayed for much of that term. We were treated to one of the most severe winters on record. The water supply in the flat was frozen most of the time. Fortunately, the flat was only a short walk from the university union building. Here we could keep warm in the evenings and wash ourselves; even manage an occasional bath before rushing back to jump into bed till morning. The Singer Gazelle remained stuck in the snow for several weeks.

During this severe weather, Geoff and I were in the middle of our third-year project that counted towards our final degree marks. The intense cold proved to be a bit of a handicap to our work programme during the evenings, but we managed to keep to the schedule. Our project made full use of the wind-tunnel to determine the drag and efficiency of various alternative air-intake configurations for jet engines mounted in the tail of passenger aircraft. Unfortunately, we had to utilize an existing, previously constructed model fuselage section, which meant we were somewhat constrained by the variation of intake configurations that we could design to fit this model. The obvious solution, such as the upper fuselage duct fitted to the Boeing 727 or the de Havilland Trident aircraft, could not be replicated on our model, so we had to settle for a simplified version. Nevertheless, the main objective of such academic projects was to demonstrate that we had the knowledge to plan and carry out the appropriate testing, followed by analysis and sensible conclusions.

At the end of the final year, after completing our project, there were the inevitable exams. The spring weather of our final term was in complete contrast to the previous term: the sun shone, the temperature rose, and I did most of my revision in the open air at the local country park. This seemed to work as, in common with most of my aeronautical colleagues, I was awarded a second-class honours degree.

In the 1960s, it was very unusual for girls to study engineering. Interestingly, there were three girls on our aeronautical course. For some reason, we had a high drop-out rate: out of the thirty-two who started the course, we were reduced to twenty-five before the final year. Sadly, two of these were girls. The one girl who stuck it out was an Australian, Sally Kitchen. She completed the course and gained an honours degree. She had great enthusiasm, and after graduation she returned to Australia with the intention of making her future in the aircraft business.

Only one student from our year managed a first-class honours degree; he stayed at the university for an extra year to gain a Master's degree. A few of us were invited to do the same, but although I had enjoyed the last three years, I was looking forward to the practical, real world of industry. It was quite sad to discover later that not many of my fellow graduates had continued with careers in the mainstream aviation industry.

During the university summer vacations, I continued with the apprenticeship course at Avro in the various technical departments. Spells in the Drawing Office, the Structures Stress Office, the Structural Test Rig, the Wind-Tunnels and finally the Flight Research and Development Department completely filled the time.

The Structural Test Rig was impressive: a whole Vulcan airframe was suspended in a massive steel frame. The Vulcan could be bent and twisted by hundreds of hydraulic rams. The rams were programmed to put loads into the wing and airframe to simulate take-offs, landings and various flight mission scenarios, including turbulence encounters. The test rig operated for twenty-four hours a day to ensure that any sign of metal fatigue damage would be revealed long before the actual aircraft reached that number of operational flying hours or missions.

At particular intervals, the fatigue cycling was paused to subject the wing to loads equivalent to its design strength limit, followed by a thorough examination of the structure for signs of damage. If damage had occurred, modifications could be developed to strengthen the local structure of aircraft in operational service.

There were three separate wind-tunnels: a low-speed tunnel that had a working section very much bigger than the one at Bristol University (one could stand up in it while setting up the model to be tested, but otherwise very similar), a transonic tunnel that could produce airflows up to and just over the speed of sound, and a supersonic tunnel. The faster the airflow in the tunnel, the greater the amount of air needed to pass through it; in consequence, the working sections had to be smaller. The air for the supersonic tunnel came from a very large, spherical pressure vessel that took many hours to pressurize. Whenever this tunnel was operated, a siren was sounded to warn of the impending ear-shattering noise that occurred as the air sped through the tunnel at supersonic speed. It could generate speeds up to Mach 3 but only for a period of ten seconds. Because the duration of the airflow through the tunnel was so short, the model under test had to be moved into as many different angles to the airflow as possible in this short time, while the aerodynamic loads were simultaneously measured. This process had to be pre-programmed before each run. At that time the supersonic – and to a large extent the transonic – tunnel was used for the design of a top-secret supersonic project that was intended to eventually replace the Vulcan. Apprentices were excluded from any involvement in this work. Much later, in the early 1970s, after this project – designated Avro 730 – was cancelled because intercontinental missiles became the preferred method of delivering nuclear bombs, this design work was declassified and we were able to see the design. It was an amazing concept conceived long before Concorde. Designed to be capable of speeds close to Mach 3, it was largely constructed from steel and titanium. It had no windscreen; forward visibility, including for landing, was provided by a TV camera and a screen on the flight-deck! Whether it would have successfully met the ambitious specification will never be known.

My final apprentice posting was to the Flight Research and Development Department, and here I was joined by my close university colleague of high-frequency legs fame, Rodney. We were allocated a desk in the office of Zbigniew Olenski, known to everyone as Bee or Olly. A wartime Polish immigrant, he was Avro's highly-respected chief aerodynamicist. Sadly, he suffered a major stroke that left him paralysed; a paraplegic. Avro continued to employ him out of respect for his accomplishments and in the hope that he would be able to contribute to the ongoing work of the department. He used an invalid car to get about. It only

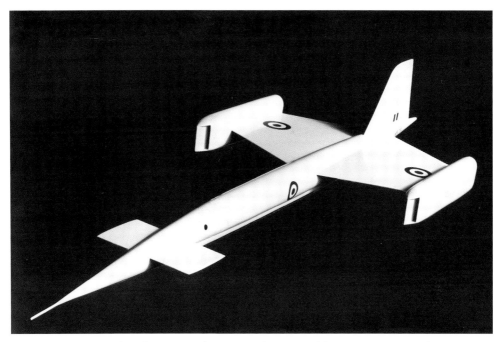

Avro 730 supersonic bomber proposal, conceived in 1955. (*Avro Heritage Museum*)

had one seat and was propelled by a very noisy two-stroke engine. By today's standards it was a crude contraption; very difficult for him to get in and out of, especially with his crutches. He was quite a character. Occasionally his crutches would jam in his car when he arrived at work; he would sit there chain-smoking while waiting for someone to help. The whole vehicle would be filled with smoke and although he could not be seen, it was obvious that he was inside and needed to be rescued.

He was, unfortunately, no longer able technically to do much useful work, but he had in his mind the concept that the movement of a pendulum hanging in an aeroplane could be used, in some way, to measure motion. He sat at his desk and, in between cigarette puffs, he would ponder over equations of motion but sadly nothing came of it.

Rodney and I were given various analysis tasks working on data from flight-testing. I found this particularly satisfying and inspiring; at this stage, it was now clear in my mind that here was where I wanted to make my career. An involvement with real aeroplanes was for me. Rodney, however, decided on a career in Sales Engineering which was based in Chadderton. Although he was one of the few of us from Bristol University who remained in the aircraft industry, sadly and surprisingly our paths never crossed again.

Chapter 4

Into the Real World

My apprenticeship seamlessly drifted into an employment contract with Avro in June 1963. It was so easy in those days; there were plenty of opportunities for graduates, particularly in the manufacturing industries. I became an aerodynamicist in the Flight Research and Development Department. At that time, Avro was primarily focused on the production and development of the Vulcan bomber, the 748 twin-turbo-propeller commuter aeroplane, and later the new Avro 780, a military transport aeroplane developed from the 748. After the delta-wing research work carried out on the Avro 707 A and B aircraft in the early 1960s to support the Vulcan design concept,[1] there was no sign of any future flight test activity that could be described as 'research'. In recognition of this reality, it was subsequently renamed the Flight Development Department. Nevertheless, there was no shortage of interesting work to be done. The Flight Test Section had two sub-sections: the Handling Section which dealt with the stability and control characteristics, and the Performance Section which was responsible for measurements of rate of climb and take-off and landing distances. I was one of five members of the Handling Section under the Section Head Ken Wood. He was, in some ways, quite easy-going but he always left one in no doubt about what needed to be done and was very approachable if you needed guidance or advice about the details of particular analysis or the priorities.

The essence of this department's work was to undertake all the flight testing and the subsequent analysis that was necessary to ensure the aeroplane was shown to meet the relevant flight requirements imposed by the certificating authority, as well as the manufacturer's design specifications. A lot of this was straightforward analysis work, but there were periods of struggling with complex stability and control equations to try to understand why some results were different from the wind-tunnel predictions. In some cases, design changes were needed to improve unacceptable or marginal characteristics.

In the case of the Vulcan, a military aeroplane, both the requirements and the specifications were specified by the Ministry of Defence (MoD). Their generic flight handling and performance requirements for all military aircraft were published in a document designated AvP 970. On top of this there were specific mission requirements, defined in the MoD 'specification' document, for each individual aircraft type for which they had placed a manufacturing contract.

The Avro 748, on the other hand, was designed for commercial airlines and, consequently, had to satisfy the civil aviation requirements. In the UK, these were specified in the British Civil Airworthiness Requirements (BCAR). These requirements were developed and published by the Air Registration Board (ARB).[2] This was a government body, initially a part of the Board of Trade and latterly the Ministry of Civil Aviation. In the USA, the civil aircraft safety requirements were published in Federal Aviation Regulations (FAR). Other countries who manufactured aircraft used either the BCAR or the FAR as the basis for their own national requirements, except for Russia (the old Soviet Union bloc) and China, which had their own uniquely military-based regulations. However,

Avro 748 prototype, G-ARAY. (*Avro Heritage Museum*)

for a commercial aircraft, it was the manufacturer who would produce the 'specification' which defined the operational capability. This would be developed having researched the needs of potential customer airlines.

A fundamental element to the analysis of flight test results was the ability to record all the necessary data. In the mid–1960s, this was fairly primitive. In the 1940s and 1950s, the standard method was to install a panel of conventional instruments (including an accurate clock) within the test aircraft. A 35mm film camera would photograph this panel at variable rates from three or four times a second to several seconds between shots, depending on the dynamic nature of the test being undertaken. The basic data of airspeed, altitude, pitch and roll attitude were easily replicated using conventional flight instruments. In addition, all the engine and system instruments on the flight-deck could also be reproduced. To record data not normally available to the pilot, such as the position of the flying controls (elevator, ailerons and rudder) the accelerations (g-forces) or the amount of control force applied by the pilot, then sensors had to be installed to measure these parameters, which were electrically connected to specially-adapted gauges on the panel. All this could result in a very comprehensive instrumentation panel. Once the films had been developed, they were sent to a team of 'film readers' who would transfer the visual readings from each instrument onto standard pro-forma sheets. These sheets were sent to the flight test analysts who would produce time-history graphs of the events to enable the assessment of relevant aircraft stability or control characteristics. One inherent problem was that many of the critical tests were often accompanied by some degree of buffeting (vibration), rendering precise reading of the instruments difficult. I can remember on numerous occasions going to the film reader's room to try to fill in some critical gaps in their readings, but very rarely could any of us improve on their interpretations.

This method was still prevalent in the 1960s, but was being superseded by the development of photographic 'trace recorders'. Avro was at the forefront of this technology. They had their own instrument laboratory largely for the manufacture and maintenance of these recorders. A light-proof metal box contained a spool of photographic paper (initially 3 inches wide, but later 6 inch ones were made) which was driven by a variable speed electric motor. A series of light points was projected onto the paper to produce a 'trace'. Each of these lights was driven across the paper by an electrical input generated from an instrument or sensor, the position across the paper being a precise record of the instrument reading. For example, airspeed and altitude were obtained from pressure sensors; control forces from strain gauges; control angles from position sensors; and pitch and roll angles directly from the electrical outputs of the attitude gyros. These recorders were a

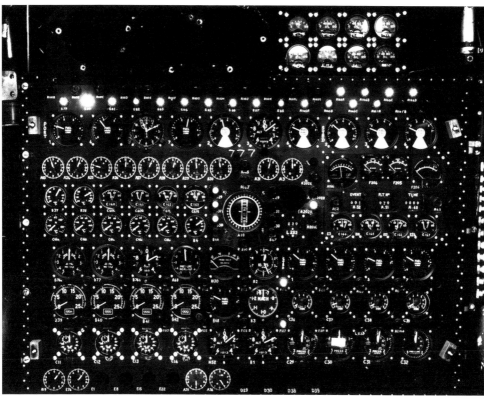

Avro Vulcan auto-observer panel, recorded with 35mm cine camera. (*Ted Hartley*)

Film-reader Valerie Twiggs beside projector, 1963. (*John McDaniel*)

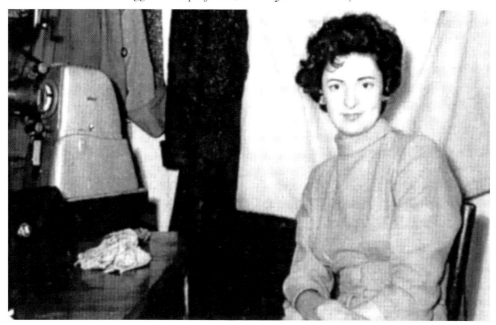

huge improvement over the film reading, giving an immediate visual relationship of the various parameters. However, they did need very careful calibration to determine the relationship between the position of the 'trace' across the paper and the parameter being recorded. The calibration graph, produced for each recorded parameter, was used to read off the value of the parameter for the specific measurement of the trace from the datum line.

The flight-test aircraft were always under great pressure to keep to a tight test programme and so the best time to accomplish the calibrations was outside normal working hours. I have atmospheric memories, which will always stay with me, of sitting in the pilot seat at the controls of a Vulcan bomber at night in a dark hangar, with all the cockpit instruments and indicator lights illuminated. I had to move the flight controls gradually in steps throughout their full range in all directions, simultaneously running the instrumentation recorder while my colleague measured the deflection of the relevant control surface at the trailing edge of the wing and the rudder. To calibrate the control force sensors, I had to pull and push the flight controls in the appropriate direction with a calibrated spring balance, again in steps, and run the recorder for each degree of applied force. These recorders were fairly sensitive to knocks, and regular calibrations were essential to ensure the results were consistent and reliable.

Although significantly better than the film reading, the method of transcribing the information from the traces into meaningful data was still time-consuming. Each individual test was marked on the paper by a dot–dash 'ident' (identification) code input by the flight test observer who simultaneously wrote in his flight-test log the details and conditions of each test. In some cases, an important specific 'event' during a test would also be identified with an 'ident' mark. The speed of the recorder paper determined the time scale for each event, which also needed an accurate calibration. The distance of each line on the recorder from the fixed datum line at the edge of the paper had to be measured by hand using an accurate ruler at each required time point. This measurement was then transformed into a value of the parameter using the calibration graph. These results were entered onto a pro-forma sheet and this data then used for the subsequent analysis appropriate to the nature of the specific test.

For any new aircraft type engaged in its certification flight-test programme, the rate at which data is being generated can easily overwhelm the capability of the analysis process if not properly planned. To appreciate the amount of testing required and analysis involved, it would be useful for the reader to have an overview of the requirements to be met, performance data to be established and other relevant aspects. These may be divided into several broad categories:

stability, controllability, stalling, performance measurement, automatic flight-control systems and system failures, as outlined below.

Stability: The aeroplane must be shown to be stable in all likely flight conditions. Stability, as the name implies, is the tendency for the aeroplane to return to its equilibrium condition following a disturbance. For instance, if the aeroplane is in steady level flight and a gust disturbs the nose upwards, the nose should tend to go back downwards towards the previous steady condition without any action by the pilot. In the test-flight conditions, this is accomplished by carefully trimming the elevator to maintain the required airspeed for the particular flight configuration. For an aeroplane with manual flying controls, where the pilot's control column is mechanically connected directly to the control surface, e.g. the elevator, there is a smaller control tab attached to the rear of the main control surface (the elevator in this example). This tab, known as a trim tab, is also controlled directly by the pilot, usually through a trim-wheel or trim switch. If the pilot moves the trim-wheel in the nose-up sense, the tab moves down which applies an aerodynamic force to move the elevator up to the position necessary to maintain the required flight condition without the pilot needing to exert any residual control force ('hands free'). Similar trim tab controls are provided for the ailerons to trim the wings level and for the rudder to trim out any sideslip.

Once trimmed in the required flight condition, the pilot will disturb the aeroplane by a small input on the appropriate control: elevator up or down, aileron to the left or right, or rudder pedals left or right and then release the control. Although each test is recorded on the instrumentation, it is visually evident whether or not the aeroplane tends to return to the equilibrium condition. Such tests must be accomplished in all take-off, landing, climb and cruise conditions and throughout the range of airspeeds appropriate to each configuration. It is not difficult to appreciate how much flight-testing is involved when one realizes that, to cover all likely cruise, take-off and landing options, there could be up to six different wing flap settings, as well as having the landing gear (undercarriage) up or down and different engine powers from throttles closed to maximum.

However, stability is always most critical with the aeroplane centre of gravity at the aft limit and for such tests other loading configurations will not be necessary. With the help of analysis and wind-tunnel test results, it is probable that the least critical configurations will be determined prior to flight testing, thus reducing the total necessary flight time.

It is generally the case that the aftermost limit of the centre of gravity permitted for the particular aeroplane type will be that where the stability requirement is only just achieved in the most critical configuration. Therefore, when designing a transport aeroplane, particular attention has to be paid to achieving good stability characteristics so that the aft centre of gravity limit can be as far aft as possible to allow maximum flexibility in the location of passengers, freight and fuel.

For aeroplanes with manual flying controls, there are a few tricks that can be employed to artificially enhance stability in certain critical conditions. For instance, it is usual that some critical cases are at low speeds with a large flap angle deployment, where the elevator angle for steady flight is towards the fully-up limit of its travel. Here, when a nose-up disturbance is made from the trimmed condition, the stability can be slightly negative; the nose tends to continue upwards when the control is released. The installation of a spring into the elevator control circuit, which takes effect at large up-elevator angles, to push against the pilot's control force (known as a 'down spring' or 'stability spring'), will help the nose to go down when the control is released.

With powered flying controls, the ability to enhance stability with electronically-computed inputs to the control circuit is almost endless, but the possibility of failures to these systems and the ensuing consequences of instability must be recognized. One fairly common example in jet transport aeroplanes is the 'Mach trim' system. When an aeroplane is flying at speeds approaching the speed of sound, the compressibility of the air can have a destabilizing effect. This usually occurs at Mach numbers above about 0.8 (80 per cent of the speed of sound), and the instability tends to increase quite quickly as the Mach number reaches the maximum limit for the aeroplane (often about 0.9 for most civil or military jet-powered transport aeroplanes). The 'Mach trim' works in a similar way to the 'down spring', but the force applied to the control circuit is tailored to match the rate at which the natural instability increases with the Mach number.

Modern aeroplanes like the Airbus types have fully fly-by-wire control. The amount of control force the pilot applies to manage the flight path is computed by software to ensure stable characteristics; the pilot may have little appreciation of the aeroplane's real aerodynamic characteristics. Such systems have to be triplicated or even quadruplicated, with cross-monitoring between each channel to identify errors in an individual channel.

Controllability: It must be possible to readily control the aeroplane in all likely flight conditions. The control forces that the pilot exerts to control or manoeuvre the aeroplane must not be excessive. This was specified in the British Civil Airworthiness Requirements (BCAR) as a maximum of 30lb (13.6kg) for the aileron (roll) control, 50lb (22.7kg) for the elevator (up and down) control, and

180lb (81.6kg) for the rudder (directional) control. The latter limit was reduced to 150lb (68kg) with the realization that some male pilots and most women could not be expected to achieve the 180lb. When carrying out the instrumentation calibrations of the rudder control forces on the 748, I was amazed how difficult it was to achieve the 180lb maximum pedal force and it was no surprise that this limit was subsequently reduced in the BCAR (and FAR) rules.

Conversely, control forces to effect manoeuvres must not be so light that precise control is impossible or that there is a danger of inadvertently applying too much control such that it may cause damage to the aeroplane. A good example of this is the amount of elevator control force necessary to initiate a rapid nose-up or nose-down change of flight path in order to, for instance, avoid an in-flight collision. The pilot's control force required must not be so low that it would be too easy to inadvertently over-stress the wings. A sudden or continuous change of direction produces a force on the object in motion, generally known as g-force. The force that gravity generates on a stationary object is defined as 1 g. It is usual to express any force as a proportion of g-force. Where there is no force, as in space, then objects are subject to zero g, known as weightlessness. Most people will have seen films of trainee astronauts experiencing weightlessness in an aeroplane. To do this, the aeroplane must perform a prolonged nose-down manoeuvre that is pre-programmed to exactly counteract the force of gravity. Manoeuvres in a nose-up sense generate forces greater than 1 g. The normal upper limit for civil transport aeroplanes is 2.5 g. In other words, the aeroplane structure must be capable of supporting the weight of two and a half times the aeroplane's maximum weight. There are safety factors in the regulations so that reaching 2.5 g does not suddenly cause the wings to break off. The elevator control force the pilot has to apply to manoeuvre the aeroplane must not be so low that this 2.5 g limit is likely to be inadvertently exceeded. Hence, the BCAR specify a minimum elevator control force of 50lb to 2.5 g; an average gradient of 33lb per g (stick force per g).

Many critical cases of controllability are where the control forces are too heavy. When the pilot wishes to change the aeroplane's configuration by, for instance, raising and lowering the wing flaps or raising and lowering the undercarriage, the resultant changes in lift and drag forces must not be so great that the pilot has to apply an elevator control force greater than the 50lb limit stated above. Similarly, the control forces resulting from increasing and decreasing the engine power or thrust must not exceed this limit. A particularly interesting case can be during the take-off. The pilot has to pull the elevator control back to raise the nose-wheel off the ground at the appropriate speed. As the aeroplane becomes airborne, very little control force should be necessary to maintain the correct initial climb speed.

The positioning of the main wheels is critical. Clearly, they must always be behind the aeroplane's centre of gravity, otherwise, when on the ground, the aeroplane would tip down onto its tail with the nose-wheel off the ground. If the main wheels are too far aft, a large elevator control force is necessary to raise the nose. As the aeroplane becomes airborne, the aeroplane's pivot point moves forward from the main wheels to the centre of lift and the pilot will then have to push the control column forward to prevent the airspeed in the initial climb becoming dangerously low. If the difference between the pull and push force is excessive, it becomes very difficult to control and can be potentially unsafe.

Almost without exception, the cases in which the control forces are critically high will be when the aeroplane's centre of gravity is at the forward limit. The low-force cases are at the aft centre of gravity limit. Again, analysis and wind-tunnel data are useful in predicting the critical cases.

As with the critical stability situations, aeroplanes with powered flying controls incorporate artificial control force supplementation into the control system mechanisms. The simplest method is often a system of springs whose effects can be varied by inputs from airspeed, wing incidence, throttle position or flap angle/landing gear configurations. Unacceptably high or low control forces can readily be eliminated in fully electronically-controlled fly-by-wire systems.

Another vital aspect of controllability testing is the determination of minimum control speeds in the event of engine failure. The basic consideration for a civil aeroplane is that an engine failure at any point, from the take-off, through the climb, cruise and landing phases, must not have catastrophic consequences. In order to achieve this, the aeroplane must remain controllable at all times should an engine fail. For reasons of commercial competitiveness, scheduled take-off and landing distances need to be as short as possible and this is highly dependent on the take-off and landing speeds being low. Clearly, these scheduled operating speeds cannot be so low that an engine failure will be uncontrollable; hence the need to establish an appropriate minimum acceptable control speed. There are three critical scenarios: the case of an engine failing during take-off while still on the ground (minimum control speed on the ground: V_{MCG}); an engine failure just after lift-off (minimum control speed in the air: V_{MCA}); and at the late stage in an approach to land when an overshoot (go-around) is necessary (minimum control speed for landing: V_{MCL}). In the latter two cases, these speeds are determined with one engine inoperative and the others at the maximum take-off power, then gently slowing the aeroplane down until the speed is reached where full rudder control (or maximum allowed control force) will just maintain straight flight. The first case, V_{MCG}, requires a more dynamic test. An engine is failed at the critical speed during the take-off run and the pilot must be able to control the lateral deviation on the runway to a maximum of 30ft (9m) before reaching the lift-off speed and continuing the take-off. These are the minimum control speeds. Of course, in a real situation, an average pilot would not be expected to fully control a sudden engine failure at these critical speeds, so the minimum take-off and landing speeds scheduled in the Aeroplane Flight Manual have margins built in to take this into account.

Stalling: When the airspeed of an aeroplane is reduced to the point where there is insufficient airflow over the wing to support the weight of the aeroplane, the wing is said to have stalled. There are two aspects of this that must be addressed during the certification testing. One is the ability to recognize and safely recover from a stall; the other is the determination of the actual speed at which the stall occurs.

There are many situations and documented cases where a pilot has inadvertently stalled an aeroplane. It is important that this should not be a catastrophic event. The pilot must immediately recognize what has happened and be able to recover the aeroplane quickly to normal safe flight, with the minimum loss of altitude. A classic stall is accompanied by a sudden drop of the nose. This is beneficial for two reasons: it is a clear indication to the pilot of a stalled situation and the nose going down encourages the airspeed to increase above the stall speed. All the pilot has to do is to rapidly apply more power to aid the acceleration so that he is able to raise the nose and minimize the inadvertent loss of altitude.

Problems arise with the aerodynamic characteristics of modern aeroplane wings. These, being optimized for fuel efficiency (minimum aerodynamic drag), have a more gradual loss of lift as the stall is approached, and consequently less of the nose-down tendency (nose-drop). Additionally, this more gradual loss of lift can lead to one wing reaching the stall slightly earlier than the other, resulting in a rolling motion. If the roll is not too severe, in combination with a small nose-drop, it can provide effective stall recognition. However, if the rolling motion is too rapid, the stall becomes impossible to control without a large height loss; an unacceptable situation.

The biggest difficulties are encountered if there is no recognizable nose-drop to warn the pilot of the stall. In the ensuing prolonged stalled situation, a considerable loss of height will rapidly result. The worst scenario is the so-called 'deep stall'. This is usually a characteristic of aeroplanes that have their tailplane mounted on top of the fin. When the wing has reached the point of being stalled, the dead air from the wing flows upwards and over the tailplane, rendering it ineffective. The pilot no longer has any means of applying the nose-down control necessary to recover the aeroplane. The usual solution introduces an artificial nose–drop, just before the danger–point is reached, by the activation of a mechanical input to the elevator control circuit known as a 'stick–pusher'.

As an additional protection against an inadvertent stall, all civil aeroplanes and many military ones must have some form of 'stall–warning' which activates at an appropriate speed margin before the stall is reached. (The BCAR states this margin should be not less than 5 per cent of the stall speed.) Pre– and early post–war aeroplane wing characteristics were such that, as the airflow began to break down as the stall was approached, a degree of buffeting occurred that was a clear indication of an impending stall. Modern wings don't exhibit this buffet and an artificial equivalent – a mechanical 'stick–shaker' – is installed, triggered to operate at the appropriate margin before the stall.

For test purposes, the classical method for reaching the stall is for the pilot to trim the aeroplane with throttles closed at a speed of about 30 per cent above the stall speed and then gently pull back on the elevator control to achieve a speed reduction of 1 knot per second. In addition to the normal stall approaches with the wings level, the characteristics of the stall when approached in turning flight with 30 degrees of bank are also assessed. Dynamic stalls, where the rate of speed reduction is increased to about 3 knots per second are also included, as well as straight stalls with a degree of power applied. The stall–handling test programme must cover all the configurations of wing flap and landing gear and, even though the critical cases are at the aft centre of gravity limit, it will consume a considerable number of flying hours.

The determination of the actual speed at which the aeroplane stalls, in all configurations, is of crucial importance to the determination of the aeroplane's performance capabilities. Just as in the case of the minimum control speeds mentioned above, for reasons of safety, all the scheduled take–off, climb and landing speeds have to be a significant fixed margin above the stall speed. Because of this, it is critical that the stall speeds are as low as possible and the importance of their careful determination during the flight tests is paramount. Again these 'performance' stalls are approached in straight flight at a speed reduction of 1 knot per second with throttles closed. The critical (highest) stall speeds are with the centre of gravity at the forward limit but, as for the stall–handling tests, a considerable amount of flight time has to be spent determining the speeds in all configurations.

Because of the critical nature of the measurement of speeds such as stalling speeds, it is important to have an accurate knowledge of any errors in the aircraft's airspeed measuring system. In the early stages of development testing, the manufacturer will fit a pitot–static probe in a location as far as possible from the aerodynamic influence of the airframe; usually ahead of the wing-tip or extending forward from the nose. The 'real' airspeed (and altitude) from this will be compared with indications of the normal flight instruments and used to determine any error.

Performance Measurements: The Aeroplane Flight Manual must contain all the information necessary to safely operate the aeroplane. This will include, for example, all the data to be able to determine the maximum weight at which the aeroplane can take off from a particular runway, considering the length of the runway, its altitude, the wind speed, air temperature and any obstacles after take-off such as high ground. All the data will be based on the premise of an engine failure at a critical point in the flight path. Similarly, landing distances will be scheduled. Before serious performance-measuring flight testing can be commenced, for the reasons outlined above, the stall speeds must be determined and preferably agreed with the Certificating Authority. The most critical and demanding performance measurements will need to be made at around the aeroplane's maximum weight. However, measurements will be made at a variety of weights to ensure that the published data covering all weights is realistic. Furthermore, measurements at high and low temperatures as well as high altitudes will be required. This will necessitate specific deployments to selected overseas airfields for tropical and cold weather trials. Noise (or lack of it) is a very important factor, and frequently noise measurements are combined with the performance take-off and landing testing.

One difficulty, often encountered in the measurement of climb performance, is the variability of the atmosphere. In a normal stable atmosphere, the temperature will reduce fairly steadily with increasing altitude at a rate of about 2 degrees C per 1,000ft. To obtain a good result, it is normal to continue a climb performance measurement for at least five or ten minutes. In that time, it is not unusual to find significant non-standard temperature variations (particularly in high-temperature environments) which can significantly upset the steady rate of climb over the period. More information on performance-testing will be found in the next chapter.

Automatic Flight Control Systems: Much has already been explained about autopilot development testing in the first chapter. The one area not covered is the automatic control of the approach to and ultimately landing on the runway. Great diligence and care are necessary in developing the autopilot control laws required for a successful automatic landing. As well as the control of the aeroplane's attitude in pitch and roll, the autopilot must successfully follow the radio direction beams, transmitted from the ground, which define the flight path down to the runway threshold. The directional path is determined by a radio beam transmitted from the far end of the runway towards which the aeroplane is heading. This, known as the 'localizer beam', is like a sheet of paper in a vertical plane, aligned with the centreline of the runway, extending out for a range of at least 20 miles. The aeroplane has a radio receiver tuned to the frequency of the localizer beam that can detect its angular distance away from this beam to the left or right (azimuth). The autopilot uses this to guide the aeroplane along the centreline of the runway. Similarly, there is a second beam transmitted from a point aligned with the runway threshold, which projects a radio beam upwards at an angle of (normally) 3 degrees above the horizontal. This is known as the 'glideslope beam' and again, it is like a sheet of paper in a horizontal plane projecting up at 3 degrees from the threshold position. The glideslope receiver in the aeroplane can detect the angular distance above or below this beam which the autopilot computer uses to control the aeroplane down the glideslope.

These two beams together are known as the Instrument Landing System (ILS). The accuracy of the localizer and glideslope beams installed at an airport which is approved for automatic landings must be assessed and qualified to a defined standard.

Obviously, it is not acceptable for an automatic landing system to fly the aeroplane straight into the runway at the threshold. A fairly gradual transition from the 3 degrees descent to a smooth touchdown is necessary. To assist this process, an accurate radio-altimeter is fitted to the aeroplane. The autopilot uses this to plot and control the 'flare' from glideslope to touchdown. In the early days, the pilot still had to control the airspeed during the approach and flare using the throttles, but when automatic control of the throttles (auto-throttles) was developed to control the airspeed, the whole process became totally 'hands-off'.

Inevitably, even the best autopilot cannot be expected to land the aeroplane smoothly at a consistent point on the runway in all reasonable conditions of aeroplane weight, centre of gravity, wind direction, turbulence and runway slope, as well as allowed variable characteristics in an airfield's ILS. Therefore, a systematic flight-test programme of approaches and landings in a representative variety of such conditions has to be accomplished to ensure, to an extremely high probability, that no landing will have damaging or catastrophic consequences. This generally will require a large number of landings to gain a good, statistically significant batch of results.

Serious consideration of automating the take-off has not yet been pursued. It is internationally accepted that the minimum visibility for commencing a take-off is 150m RVR (Runway Visual Range). This will normally be sufficient for the actual take-off to be accomplished without any automatic guidance. Once airborne, normal instrument flying or engagement of the autopilot are acceptable safe options and in the event of any unforeseen need to land back at the airfield, the automatic landing system is available.

As an extension of this scenario, modern aeroplanes have comprehensive Flight Management Systems (FMS). The whole detailed route, in three dimensions, from the point of becoming airborne, through the climb phase, the cruise, the descent into the landing area and, if necessary, an automatic landing, can be pre-programmed into the FMS for the autopilot to follow without any pilot intervention.

System Failure Testing: The complexity of aeroplanes inevitably leads to the possibility of system failures which have to be considered. One simple example is the failure of the undercarriage to extend for landing. In every commercial and military aeroplane, there is always at least one alternative method for getting the wheels down should the normal system fail. Almost invariably, the primary means uses a system driven by an electric motor or hydraulic rams. In the event that this fails to function, the simplest back-up system relies on gravity. For this, the undercarriage compartment doors and the up-locks must first be released. In some cases, this can be accomplished manually by the flight crew operating emergency levers, generally located under panels in the cabin floor. This should allow the doors and undercarriage to drop down. So far so good, but to be successful, the undercarriage down-locks have to engage, otherwise, on touchdown, the undercarriage will simply be pushed back up into its compartment and the aeroplane will 'land' on its fuselage. When this test is undertaken in the hangar with the aeroplane suspended on jacks, the inertia of the falling undercarriage will firmly drop it into the locks. In the air, however, there are aerodynamic forces at work and possibly some small structural distortions in the landing gear bay, either or both of which may contrive to slow down the free-falling action, hence the importance of checking it in flight. Sometimes it may be necessary to undertake a little manoeuvring in pitch to get the locks to engage.

The more complex the aeroplane, the greater the possibilities for system failures to occur. Even more important than the undercarriage are the flying controls. In the case of simple aeroplanes with direct mechanical control linkages from the pilot to the control surfaces, such as the elevator, with careful design to avoid undue wear or potential for jamming, these were considered to have sufficient in-built integrity that failure was very unlikely. However, after the much later introduction of numerical safety assessment requirements for systems, it was determined that failures of mechanical systems had to be considered (this 'safety of systems' is dealt with in more detail in Chapter 12). This required duplicated control linkages from the pilot to the control surfaces which, in the event of a failure or jam of one linkage, the two halves could be disconnected and control maintained through the un-jammed linkage. Powered flying controls using either electric or hydraulic activation are a different matter. Hydraulic systems can leak; electrical generators and their switching can fail. For multi-engine aeroplanes, it is usual for each engine to drive its own separate generator and hydraulic pump. The way the supply from each engine is interfaced is crucial to the ability to cope with an engine failure as well as various likely single failures in a system. The procedures in the event of failures must be assessed in flight to ensure that the pilot can always maintain control of the aeroplane.

Twin-engine aeroplanes have more potential for total loss of hydraulic or electrical power than three- or four-engine ones. Some additional source of power will usually have to be provided. This is often an Auxiliary Power Unit (APU): a small turbine (jet) engine, mounted in the aft fuselage, that normally supplies power to the aeroplane while on the ground, before the normal engines have been started. However, in order for the APU to be available for emergency purposes, it has to be possible to start and operate it in flight when necessary. This must be demonstrated during flight-testing. Another source of emergency power, often installed, is a Ram Air Turbine (RAT): an electrical generator and/ or a hydraulic pump driven by a small propeller which, when needed, can be lowered into the airstream. Again, the means of lowering and the procedures for successfully distributing its power into the failed systems have to be demonstrated in flight.

With pressurization systems, pneumatic systems, flight instrument systems and all manner of automatic flight control systems to add to the list of potentially serious failure cases to be considered, it is clear that considerable flight test time has to be allocated to such work.

★   ★   ★

To return to my work at Woodford in the Handling Section, the significant projects under way at that time were the early development testing of the military transport version of the Avro 748, designated the 780 (it was also known as the 748MF and later named the Andover), the Vulcan automatic landing, and the Vulcan Skybolt missile programme. All needed comprehensive testing and so there was plenty of analysis work to be done.

Avro 748MF prototype G-ARRV, showing rear cargo door open in flight. (*Avro Heritage Museum*)

Avro Vulcan with Skybolt missiles, 1964. (*Avro Heritage Museum*)

The Skybolt was an American-designed air-launched intercontinental ballistic missile and was intended to be the next generation of nuclear deterrent. The UK government had concluded that it was easier and cheaper to purchase these missiles rather than embark on a similar parallel UK-designed alternative. The obvious choice of the aeroplane from which to launch these missiles was the Vulcan. One was fitted beneath each wing, which required some local structural strengthening. (Many years later in the Falklands War, this wing-strengthening feature enabled the air-to-air defensive Sidewinder missiles to be carried under each wing.) Although most of the development testing had been completed, there was ongoing analysis and some testing. However, just before the test Vulcan was about to be flown to the USA for the live launching trials, the US government made the sudden decision to cancel the Skybolt programme altogether. The submarine-launched Polaris missile was being developed. It was clearly a more promising nuclear delivery option and the UK government followed the US decision.

The Vulcan auto-land was especially interesting. This operational requirement was conceived before it became a possibility in civil aeroplanes. The later automatic landing systems used the localizer and glideslope radio beams for guidance. At the time the Vulcan auto-land was being developed, it was not thought that the capability of the current state-of-the-art localizer beams, being sited at the far end of the runway, would give an accurate enough left/right guidance close to the landing-point. The glideslope transmitter was sited close to the touch-down point and could provide the necessary accuracy in the up/down sense when combined with a radio-altimeter. The Avro/MoD solution was to replace the localizer guidance with a 'leader-cable'. This was a long transmitting cable laid underground along the extended centreline of the runway out from its start. It was over 1 mile in length. A leader-cable was installed at the Royal Aircraft Establishment (RAE) airfield at Bedford for the trials. The Vulcan used the localizer beam guidance until it was about 300ft above the ground when it transferred to the leader-cable signal. This proved successful, but it would have required a major excavation exercise to be carried out at each RAF airfield that required the Vulcans to land in all weathers. It was partly due to the extensive cost of equipping all necessary Bomber Command runways with this that ultimately led to the UK government reconsidering the operational necessity of all-weather landing, and the project was later cancelled.

The analysis of the auto-land results gave me my first introduction to autopilot control laws and the way the parameters could be varied to optimize the guidance performance. Derek Bentley was the senior engineer in charge of the analysis. He was an extremely clever engineer with a degree in mathematics and was always ready to share his knowledge and thinking, from which I learned a lot. Over the next years, we worked together on many projects and developed a close working and social relationship.

The Avro 780 (748MF/Andover) was a military transport aeroplane designed for operation from short and poorly-surfaced runways. It was a significant development of the Avro 748 and was designed to operate at higher weights. It had more powerful Rolls-Royce Dart turbo-propeller engines and a large rear loading door and ramp in the tail. Because it had larger diameter propellers, the landing gear needed to be longer than that of the 748. This meant that the rear fuselage was a long way from the ground. A significant innovation designed to overcome this problem was the ability to 'kneel' the main landing gear on the ground so that the rear ramp could be lowered suitably close to the ground for military vehicles to access.

Although a military aeroplane, it was based on a civil type and it was decided mutually between Avro and the MoD that the civil BCAR would be a more suitable certification code than the rather outdated transport section of the MoD AvP 970 requirements. The prototype Avro 780 was converted from one of the 748 prototypes, G-ARRV, and first flew in December 1963.

At this time, following the government policy to rationalize the UK aviation industry, Avro became a part of Hawker Siddeley Aviation along with other notable aviation companies such as de Havilland, Hawker and Armstrong Whitworth. From then on, the 'Avro' designation was replaced with 'Hawker Siddeley';

Avro 748MF demonstrating kneeling undercarriage. (*Avro Heritage Museum*)

not such a 'roll-off-the-tongue' name. Because of the predominance of South American operators who had bought the 748, it was affectionately known as 'el Avro'.[3] 'El Hawker Siddeley' did not have the same 'ring' and never caught on! Most of us, out of loyalty, continued referring to the company and the aeroplanes as 'Avro' for a few years (as I have done in this book), before it eventually became frowned upon by the management.

The initial flight testing of any new type is invariably carried out with the aeroplane loaded at a mid-centre-of-gravity position until the predicted flight characteristics are shown to be fairly accurate and systems are working as intended. This is followed by some systematic assessments of forward c.g. controllability cases. If all these initial results look to be acceptable and close to expectations, then the aeroplane will be loaded to begin the investigation of the more critical aft c.g. cases.

I clearly remember the first aft c.g. flight of the Andover. Our flight test office had windows overlooking the airfield and we regularly watched the take-offs, particularly anything special. On this occasion, we were all at the windows. The aeroplane accelerated rapidly down the runway and leaped into the air. Soon it was obvious that something was wrong. The aeroplane was climbing at an unrealistically steep attitude like a jet fighter, and even with full power the airspeed was obviously reducing dangerously towards a stall. Chief Test Pilot Jimmy Harrison was at the controls and we watched him roll the aeroplane rapidly to about 60 degrees of bank to get the nose down and then reduce the power. Control was obviously regained, and the wings were brought back to level, followed by a circuit of the airfield back to land. The problem was simple. Despite the information from the wind-tunnel tests and analysis from earlier flights, there was not enough available down elevator to control the airspeed with maximum take-off power at the aft c.g. limit. It was my first direct experience of the potential dangers of flight testing. A design change to increase the maximum down elevator angle and a slight reduction in the aft c.g. limit were the solutions, but it inevitably introduced a significant delay into the programme.

Actively involved in the flight testing were three regular flight test observers: Ted Hartley, who was primarily responsible for the military work on the Vulcan; Mike Turner who did nearly all the flight test observer duties on the 780 trials; and Andrew McClymont. Mike was extremely experienced, very capable and highly thought of in the department. To his enormous credit, he had achieved this with little in the way of formal technical qualifications. For the performance testing, Mike was assisted by Andrew McClymont from the Performance Section who often stood in for some of the handling testing.

Around this time, I was beginning to become interested in the idea of taking an active part in the flying programme. As soon as I approached Department Head John McDaniel with this suggestion, he took it very seriously as everyone was well aware of Mike Turner's workload. John had to clear it with my section head Ken Wood to make sure he was happy that I could be spared from some of the analysis work. Ken readily agreed as he was not one to stand in the way of his staff's ambitions and he saw the net benefit to the overall performance of the department.

In September 1964, I took my place at the flight test observer's position in the prototype 780, G-ARRV. I was situated about halfway down the fuselage, facing a panel with numerous flight instruments and gauges replicating some of the pilot's flight instruments and other information such as the normal acceleration (g in the vertical direction). There was a small console on which to write the test log and keep the flight programme as well as all the necessary graphs and data from which I had to determine the correct configurations and speeds for each test. On the console was a switch to start and stop the recorder, a selector to vary the recorder speed as necessary and a button to inject the unique dot-dash 'idents' for each individual test. Sensibly, for my first few flights, Mike Turner was there to watch over me, making sure everything went according to plan. Flying time is expensive and not to be wasted. Chief Test Pilot Jimmy Harrison was the pilot; another reason not to embarrass myself. Mike's first instruction was always to run the recorders while the pilot went through his full travel control checks before taxiing, thus having a record on the instrumentation to confirm the correct amount of control travel. With a bit of coaching from Mike, all went well during the tests and after two flights, I was on my own.

The flight-test programme was produced by the Flight Test Section and was organized to cover all the cases specified in the relevant Airworthiness Requirements which, in the case of the Andover, were the BCAR of the ARB. It was always prudent to perform the least critical testing early in the programme to minimize the chances of encountering unexpected problems that may be potentially dangerous. The results of the wind-tunnel testing and aerodynamic calculations were continually used to predict the critical areas and the earlier flight test results were compared with these predictions before exploring the most difficult areas. Section Head Ken Wood developed a master programme to take us through all phases of the flight testing to demonstrate compliance with all the requirements. This programme also allowed for any possible development work where the aeroplane may not be able to meet a specific requirement and may need to be modified in some way. As the flight test observer, it was usually me who would produce a programme for each flight based on Ken Wood's master plan.

Ken Wood and I would always discuss these detailed proposed programmes with the project test pilot and department head to make sure they were happy and we were not stepping too far into untested territory. The aeroplane was loaded to the required weight and centre of gravity for the test programme with the fuel load and any additional ballast installed in the cabin as calculated by the weights engineer.

I soon recognized that the amount of data to which I had to have rapid access needed organizing. There were the graphs of stalling speed against weight; one for each scheduled flap setting. The 748 and the Andover had many different specified flap angles. Apart from the normal cruising case with the flaps fully retracted (known as the 'clean' configuration), there were three alternative take-off flap angles of 7.5 degrees, 15 degrees and 22.5 degrees, and for landing, 27.5 degrees. The 22.5 degrees setting was also used for some landing situations. The Andover had an additional short landing setting of 30 degrees. Many of the test conditions were necessarily at airspeeds that were a specific factor above the scheduled stalling speed; for instance, the longitudinal stability in the approach to land configuration was assessed at 1.3 times the stall speed. There was the maximum operating speed and maximum allowed speeds for each flap angle and for when the undercarriage was down. There was the weight and c.g. limit chart and the altitude limits.

I kept all these charts and information on a clip-board with 'tabs' so that I could turn up the required information immediately. I also had a stop-watch fixed to the clip-board which was easier to use than the one mounted on the observer's instrument panel. It was one I had acquired from an ex-government surplus shop; really good quality and with a split-second hand that could be stopped independently of the main second hand to allow two separate timings of actions during one event.

It was necessary to know the fuel contents at the beginning of the flight to calculate the actual aeroplane weight, from the amount of fuel consumed, at the start of each specific test. Fuel contents gauges could be calibrated in gallons, US gallons, litres, pounds or kilogrammes, depending on the preference of the particular customer. Our test aeroplanes usually used gallons or pounds. If they were in gallons, one had to know the specific gravity of the kerosene. The 748 and Andovers also used a mixture of water and methanol from a separate tank to increase the power of the RR Dart engines on take-off. The methanol increased the combustive power of the kerosene, but the addition of water was necessary to keep the temperature of the turbine blades within limits. The amount of water-methanol used also had to be taken into account. Before the days of hand-held calculators, I carried my slide rule to do those calculations which were beyond my mental arithmetic capability.

As a flight test observer, I also needed to be familiar with some general emergency procedures and situations. The MoD had an Aeromedical and Safety Training School at the Aircraft and Armament Experimental Establishment (A&AEE), Boscombe Down. It was usual for the manufacturers to send their test aircrew bi-annually to this facility. I was duly despatched on the next available course which was, conveniently, after a couple of my test flights. There were various categories of the course and the one most suited to our needs was Category 'B', which was for aircraft flying up to an altitude of 50,000ft with a cabin altitude not exceeding 25,000ft (50,000ft was well in excess of my immediate needs, but experience of cabin altitudes above the normal 10,000ft maximum in a pressurized aeroplane was necessary in cases of pressurization failure, either inadvertent or deliberate for test purposes).

Apart from some useful lectures covering human physiology and emergency equipment, there were four practical aspects of the course: pressure breathing, hypoxia experience, parachute training and wet dinghy drills.

The pressure breathing involved wearing a helmet and mask which was supplied with oxygen under pressure to ensure the wearer would get sufficient oxygen into the lungs in situations where local air pressure was so low that even 100 per cent oxygen was insufficient to provide the lungs with enough to keep the brain and body functioning (above about 40,000ft altitude). The mask had to be tight-fitting so that the pressurized air did not leak out. One had to learn to forcibly exhale. I found this to be a very unnatural thing to do and, coupled with the uncomfortably tight mask, I was pleased when it was all over. It was a useful lesson for people likely to be flying in high-performance military jets, but it had little relevance for transport-type aeroplanes. However, it was an interesting educational experience for me at that time.

The hypoxia exposure was more relevant. We were all (about ten of us, including two instructors) seated inside a sealed pressure compartment wearing oxygen masks. The air was extracted until the internal altitude was about 25,000ft. We each had a clip-board and pencil and the instructor told us to write down the number 200 at the top of the page. We were then asked to remove our masks and carry out a mental arithmetic test: to subtract seven from the number, write down the answer and again subtract seven, continuing for as long as we could. At the same time, the instructor emphasized that we should each try to remember any symptoms of which we became aware. The two instructors in with us obviously kept their masks on to monitor our progress.

When satisfied that we were all comfortably breathing, the chief instructor said: 'OK, unclip your masks and start the subtractions. Don't worry, we will replace your masks as soon as we see you in difficulty.'

I started well and was some way down my sheet. Then I began noticing a tingling in my fingertips (one of the usual symptoms), but I continued trying to concentrate on the mental arithmetic. I was vaguely aware that some people's masks were being replaced and before I was reconnected with the oxygen, I knew that I was noticeably failing. As soon as the instructors noted anyone whose writing had degenerated into complete gibberish or had stopped writing altogether, they would rapidly replace the subject's mask. Being able to recognize one's own symptoms could be critical in cases of inadvertent loss of cabin pressure.

The parachute training did not involve actual parachuting; too much potential for serious injury. Firstly, we learned how to strap on the parachute and operate the rip-cord to open the canopy. We were then asked to jump off a 3ft-high deck and were taught how to roll sideways on landing so that the side of our knees, our bottom and then shoulder progressively absorbed the impact. The other feature common to most military aircraft was that the seat 'cushion' to which you were strapped was a one-man dinghy and survival pack. After bailing out of the aircraft and with the parachute deployed (hopefully!), as you estimated the ground to be a few hundred feet away, you released the clips to this dinghy-pack. It would drop away but it was still tethered to the harness on a long line (a lanyard). This had two benefits: firstly, it reduced the weight to be absorbed on landing and, as it touched the ground or sea, you knew it was the moment to prepare for the impact.

The dinghy drills involved jumping into a tank of water wearing a flying suit, with the dinghy-pack attached to your harness by its lanyard. This was appropriate for military fast-jet aircrew who sat on the dinghy-pack cushion in the cockpit, often on an ejector seat. Transport aeroplanes invariably had multi-person dinghies. Nevertheless, the principles were the same: get out of the parachute harness straps, pull the dinghy-pack towards you, get it inflated immediately, climb into it as soon as possible and bail out the water to get as dry as possible as quickly as possible.

Over the next five months, I did forty-five flights with a total of about seventy-seven flying hours in G-ARRV and the initial development flight testing came to an end. All the data necessary to delineate which areas would need some design change in the production version had been gathered.

Many of the tests had involved some degree of sudden and fairly violent manoeuvring; in particular rate of roll, stick force per g, and dynamic stall tests. Participating in test flying was not going to be a part of a successful career if motion sickness was a problem. Until you tried it, you could never be sure just how well your stomach would cope with this sort of activity. Sitting at a console, concentrating on the instrument panel and writing data into the log while the aeroplane was being subjected to manoeuvres in various directions is a testing scenario. Having a window alongside the test panel was some help, even though it only offered a vague peripheral vision of the outside world during the testing. The ground maintenance engineers always made sure there were sick-bags available at the observer's station; they had plenty of experience of cleaning up the mess inside their aeroplane. Early in my flying, I did find myself having to make use of this facility on a couple of occasions, but I never let it get the better of me during the tests and the understanding ground engineers cleared the bags away without any fuss.

Only once did the test pilot notice the ground engineer removing the sick-bag after the flight. 'I didn't realize you were having a problem. You should have let me know and we would have paused the testing.'

Stoically, I replied: 'It was not a big problem and I didn't wish to interrupt the programme.' I was determined not to let it get the better of me.

It was during this period that I started to take an interest in car rally navigating, encouraged by Derek Bentley. Derek was an experienced navigator himself, but had decided to give it up and try a bit of driving. He had a standard 850cc Mini, so he was never going to be competitive against the Mini Cooper Ss or Ford Lotus-Cortinas of the time. However, he just wanted to enjoy the rally-driving experience, so he needed an enthusiastic capable navigator. This is where I came in; plenty of enthusiasm but little or no real experience. Derek took me on a few practice runs to see how I managed. He was both patient and encouraging and things went well enough for Derek to enter us in a local car rally. These rallies took place at night; usually starting around midnight when the country roads were fairly empty. We did quite well, considering my inexperience and the car's lack of performance. Derek certainly got the most out of the car and it was only the uphill sections where we were woefully disadvantaged. I made a couple of mistakes, but between us we corrected ourselves fairly swiftly. Head down studying the route and giving instructions was another challenge to any potential motion sickness tendency, but I never found it a problem. I am sure that this and the flight-testing activities combined to rapidly improve the strength of my stomach.

The next phase of the flight testing awaited the completion of the first production aeroplane on which all the proper certification testing would be undertaken. This would be some ten months away.

Notes

1. Fildes, *The Avro Type 698 Vulcan: The Secrets Behind its Design and Development* (2012).
2. Chaplin, *Safety Regulation: The First 100 Years* (2011).
3. Blackman, *Test Pilot: My Extraordinary Life in Flight* (2009), p.134.

Chapter 5

No Time for Boredom

While waiting for the production Andover to be ready for flight testing, I was busy completing analysis and report-writing from the G-ARRV and Vulcan flight tests. However, my new-found experience and capability as a flight test observer was not going to be wasted. The next project in which I became involved was very different to the 780: the Avro Shackleton.

The Shackleton was a maritime reconnaissance development of the wartime Avro Lancaster bomber. The maximum weight was increased and the Merlin engines were replaced with more powerful Rolls-Royce Griffon engines, each with contra-rotating twin propellers. It still had the forward bomb aimer's window in the nose but internally it was filled with maritime surveillance

Avro Shackleton Mk 3. (*Avro Heritage Museum*)

and rescue equipment, such as sonar-buoys that were dropped into the sea for detecting underwater submarines, depth-charge launchers and often a life-raft in the bomb-bay that could be dropped for rescue purposes. Initially, the Mk 1 and 2 versions retained the tail-wheel configuration of the Lancaster, but the Mk 3 was modified to have a more up-to-date steerable nose-wheel configuration. The Hawker Siddeley Nimrod, a development of the de Havilland Comet, had been chosen by the MoD to replace the Shackletons, but it was then some two years away from its first flight and many more years still from entering service. In the meantime, to fill the operational gap, the Shackleton was being asked to carry more and more equipment and operate over longer distances. More equipment and more fuel inevitably meant that a further increase in maximum take-off weight was necessary.

The development of the Griffon engines had reached its limit and they were no longer powerful enough to get the required extra weight airborne from the available runways. The solution, conceived by the Avro design engineers, was to provide the extra power for take-off by installing a pair of Viper jet engines; one in the rear of each outboard Griffon nacelle. It was known as the Viper-Shackleton. These engines had been in use for some time in the Jet Provost RAF training aeroplane and later in the first version of the de Havilland (now HS) 125 business jet. Even by the standards of the time, these were very noisy and quite thirsty engines but, in this Shackleton application, they were only used for the take-off. Being jet turbine engines, their normal fuel was aviation kerosene; the Griffons used aviation gasoline (petrol). To fit separate tanks would have been overly complex and a further addition to the maximum weight. The Viper fuel system was, therefore, modified to cope with the higher explosive nature of gasoline. The only downside to this was that the fuel to the Viper and the Griffon in the outboard nacelles had a common supply line. Hence, the critical single failure case was the interruption of this common fuel supply during take-off, which would result in both engines stopping concurrently. As described earlier, it is essential to be able to control an engine failure at the most critical point on the take-off and safely continue into the air.

The project test pilot was Bill Else: a very capable and amiable pilot who had joined the Woodford test team from Armstrong Whitworth when that company closed in the early 1960s. He had flown Hawker Typhoons on ground-attack missions in France towards the end of the war. During a mere one week in early April 1965, and in about twelve flying hours on Viper-Shackleton WR973, we accomplished all the testing necessary to establish that there were no adverse flight-handling characteristics associated with these modifications and the stalling

Avro Viper–Shackleton development aircraft WR973. (*Avro Heritage Museum*)

Avro Viper–Shackleton showing engine installation and Viper engine air–intake extended for take–off. (*Avro Heritage Museum*)

speeds were not affected. We were then ready to attempt the critical engine failure take-off tests.

There was no doubt that this would be a potentially dangerous test and, for optimum safety, we used the very long and wide runway at the Royal Aircraft Experimental Establishment at Bedford. This turned out to be a good decision. The width of the runway was available when we needed it to accommodate some larger than expected lateral deviations that occurred after the simulated engine failures on take-off. On another of the test situations, the extra length of the runway enabled Bill to abort the take-off after failing the two engines when it became apparent that the aeroplane might not get airborne at all. Some fine-tuning of the scheduled take-off safety speeds was required before we tried again later the same day.

In Bill's own words: 'If I can get it into the air for long enough to retract the undercarriage, with the reduction of drag, it will climb OK. Sometimes it just needs a lump in the runway at the right point to help it into the air!'

This emphasized just how critical the double engine failure was at the take-off weights we were trying to approve. This situation would be unacceptable for a civil aeroplane, but the military were more ready to compromise. We did a bit more analysis of the test results and added a further knot to the take-off speeds for a little extra comfort. It was then time to begin the flight test measurements of the actual performance: rate of climb, take-off and landing distances, which would be published in the Aeroplane Flight Manual. Andrew McClymont took over from me for this phase of testing. Not long afterwards, Andrew left Avro to join the Air Registration Board (the forerunner of the Civil Aviation Authority Safety Regulation Group) as a performance flight test engineer. This was a big loss for Avro, but we had Bill Horsley there ready and able to step into his shoes.

The installation of the Viper engine was a successful modification and a small number of Mk 3 Shackletons were converted and operated by Coastal Command. However, the increased operating weight had a serious detrimental impact on the fatigue life of the wing spars. Although a frequent inspection programme was initiated to look for cracks developing in the spar and repairs implemented as far as possible, it was a losing battle and these Viper-Shackletons had to be withdrawn from service some time before the replacement Nimrods became available.

In between specific flight test projects, the 748s were coming off the production line. They needed to go through the production flight test schedule before being issued with their Certificates of Airworthiness. I was able to fit in some of these tests in quiet periods. One of these aeroplanes was the first military-registered 748 for the Queen's Flight. For some reason, the RAF had given it the designation of an 'Andover CC Mk 2' which was confusingly close to their 'Andover C Mk 1' designation for the military freighter, the HS 780.

The next project allocated to me was the conversion of the Armstrong Whitworth Argosy for use as an in-flight refuelling tanker. I seemed to be spending a lot of time with the previous generation's types, but I had no complaint; they were always interesting and sometimes challenging. Armstrong Whitworth had designed and built the Argosy, but the company was no longer in business and Avro had inherited the responsibility for the type. It first flew in 1959 and was used as a freighter in both civil and military guises. It had the wing mounted on top of the fuselage and was powered by four Rolls-Royce Dart turbo-propeller engines; an earlier version of that used in the 748. The tail was mounted on twin booms extending behind the wings, so that doors could be installed in the rear of the fuselage for direct access from the ground. The MoD needed to extend the range of the Argosy to enable it to deploy military payloads over longer distances. The plan was to use one Argosy as a tanker to refuel another Argosy that was en route to its destination. They were affectionately known as 'buddy tankers'. All such Argosies were to be fitted with a receiving probe mounted above the cockpit. Those being used as tankers were fitted with a removable hose unit attached externally to the port rear fuselage, from which the refuelling hose and drogue would be deployed. Bill Else and Eric Franklin shared the test flying. Like Bill, Eric had joined us at Woodford from Armstrong Whitworth where he had been chief test pilot; both had plenty of experience flying the Argosy. Eric, like Bill, was very capable and approachable and he had also seen wartime action as a Halifax bomber pilot on raids over Germany.

It only took about seven flying hours to establish that the addition of the nose probe and the external hose unit fitted to the rear fuselage had no significant detrimental effect on the handling qualities. We then had to establish that a receiving Argosy could readily couple with the hose extended behind the tanker. It had to be established that the airflow around the tanker did not cause any instability at the end of the hose, which would make it difficult or impossible for the receiving pilot to insert his probe into the drogue at the end of the hose. With Bill flying the receiver and Eric the tanker, it took a couple of flights to develop a technique and find the optimum speed, with a further two flights to fully demonstrate the viability of the concept. Sadly, despite the success, the MoD subsequently decided that the idea was not worth pursuing.

While I was involved with the Argosy project in late 1965, the production Andover, XS 595, had made its first flight in July and everything was progressing well with Chief Test Pilot Jimmy Harrison and Mike Turner as the flight test

Armstrong Whitworth Argosy XN814 used for developing testing of in-flight refuelling system. (*Avro Heritage Museum*)

observer. It was not until the second production aeroplane XS 594 joined the test programme in March 1966 that my services as flight test observer were back in serious demand. Tony Blackman was the primary project pilot; I was getting to know him very well and we became a good team. We were doing all the really interesting handling testing and my stomach was standing up well, despite there being no window directly adjacent to my console. For the riskiest test work, it was normal policy to wear parachutes. However, baling out in an emergency could be quite difficult through the normal 748 entrance door but, in the case of the Andover, the best option was to open the rear cargo door and leave from the ramp. For some unexplained reason, I had a vision of running down the fuselage and leaping into space from the ramp while shouting 'Geronimo!' It never happened.

During the next seven months, I did forty-seven hours on the XS 594, almost all with Tony. In the middle of this we invited the ARB to make a brief handling assessment of some critical areas. Because it was a military type this was not strictly necessary, but since the MoD had decided to adopt the BCAR instead of AvP 970 as their criteria for new military transport aeroplanes which were based on a civil type, it seemed like a good idea. Gordon Corps was the test pilot and Keith Perrin their flight test engineer. I had not met either before but unbeknown to me at that time, they would both become close colleagues in future years. Gordon and Keith had a couple of flights looking at the usual critical areas of stability, stall handling, minimum control speeds and controllability stick forces but although some results were marginal, they found nothing on which to report adversely. The ARB was not asked to produce a formal report. The final compliance reports would be submitted by HSA at Woodford to the MoD for their review and, hopefully, their acceptance.

This period of testing also included the optimization of autopilot control laws. The Andover was fitted with the Smiths Industries SEP 2 autopilot, similar to that in normal 748s. We at HSA were primarily in control of this development, although we were closely assisted by the Smiths representatives. It was much later that I discovered the more usual approach adopted by other manufacturers was for the autopilot manufacturer to take the lead, but with the aeroplane manufacturer in close co-operation. It was my first experience of selecting and assessing the variable parameters of the autopilot control laws using the autopilot test-box. I gained a lot of valuable knowledge; especially the effect that changes to the sensitivity of the autopilot control servos had on controlling various aeroplane flight parameters such as height, airspeed, pitch, roll or yaw attitudes and rates of change of attitudes, as well as the effect of introducing small time delays into the autopilot computations.

In the summer of 1966, some of us Woodford flight crew were despatched to RAF Mountbatten at Plymouth for a sea dinghy training exercise. The test flying I was involved in at the time was on the HS 748 and the Andover military freighter, both of which had multi-person dinghies. This training exercise was with single-man dinghies as used in military jet aircraft. The dinghy-pack was effectively the seat cushion and was strapped to one's bottom. I was not sure why I was chosen to be on this course; I was not involved as a flight test observer in any of the Vulcan flight test programmes. However, it would turn out to be a valuable experience.

Firstly, there was some classroom tuition. In the event of an ejection or bail-out, you would deploy the parachute and the procedure was to release the dinghy-pack when the ground (or sea) was approaching. It was still attached to you with a lanyard of about 15ft in length and as soon as it hit the water, you were supposed to release the parachute so that it would not drag you through the water after landing. However, for this exercise we would not have a parachute so we had only to deal with the dinghy-pack.

When it came to the actual dinghy drills, the instructor particularly emphasized: 'The important criteria for survival are to get the dinghy inflated quickly, climb in and start bailing out the water to create the driest environment possible. Being dry will give your body the best chance of keeping warm. You can then start assessing your survival equipment.'

After some lunch, about which most people were a little apprehensive (but not over-confident me!), we were taken out to the middle of Plymouth Bay on an RAF Air-Sea Rescue launch. There were ten of us in all; a mixture from the military services and the aircraft constructors. We had changed into flying overalls with just our underpants and a thick woolly vest. We assembled on the stern deck, each wearing a life-jacket and clutching a single-man dinghy-pack attached to our harness with the standard lanyard. The launch was about the size of a Motor Torpedo Boat (MTB), and the deck was some considerable height above the sea. The instructor gave us each a number and lined us up in order. I was number three.

'Inflate your life-jackets.' When he was sure that all our jackets were properly inflated, he continued: 'When I call your number, throw your dinghy-pack off the rear deck and follow it into the water.'

Summer it might have been, but the shock of being dunked in the cold water was the first surprise. The sea state was much rougher than it looked from the deck. I found the lanyard and pulled the dinghy-pack towards me. I grabbed the inflation handle and yanked it as strongly as I could. After the second attempt, thankfully it started to inflate. It was not at all helpful to find that it had settled upside down on its canopy. A great deal of effort was needed to right it in the unhelpful sea conditions. Eventually I pulled it the right way up and, with what seemed like my last remaining energy, I managed to struggle on board. By this time there was a lot of water in there with me. With the instructor's words ringing in my ears, I found the bailer and set to work. It took a long time to make much impression; I was beginning to suspect a leak but after more effort I seemed to be winning. Next came the onset of a distinct feeling of nausea. Despite having one of the strongest stomachs in the business, I was very sick. I found the sponge and continued to try my best to get water out. I started to look around and was soon aware that none of the other dinghies were in sight. Sensibly, I had left my glasses on board the launch and thought it may simply be an eyesight problem. The dinghy was bobbing about in all directions to the waves, which was not supposed to happen as there was a sea-anchor (a drogue attached to the dinghy with a line) to keep the dinghy pointing into the waves. Still feeling very sick, I leaned over the side and tried to find the sea-anchor line in case it had become fouled on some part of the dinghy. I found the line but there was nothing attached to it. No wonder I had been so sick. This was also the explanation for having drifted away from the others.

After some time, I heard the engine of the RAF launch and soon its welcome shape came into sight. I was hoisted on board, had a warm shower, got into dry clothes and the nausea was a distant memory. My colleagues made jokes about enjoying it so much I didn't want to come back, but they did manage to express some sympathy. I learned that, because I had drifted quite far from the others, the launch had difficulty in finding me. I was, by some considerable duration, the longest in the sea.

A very educational experience, but one I had no desire to repeat ever again.

★   ★   ★

Away from work, my interest in car rallying, following Derek Bentley's encouragement, was increasing. Avro had a thriving motor club: the Avro (Woodford) Motor Club. The majority of members were members because they were able to borrow car maintenance tools from the club's extensive collection. There was, however, a good nucleus of competitive-minded members who entered various local treasure hunts, rallies and other such events. Derek had inspired a number of these to join him in organizing a night rally to be held in the northern Welsh hills. It was named the Avro (Woodford) Moorland Rally. Over the next few years it became a classic event in the rally calendar of the local clubs. Some members were a little too old to be serious competitors but one in particular, Harry Wood, who was a store keeper on the assembly line, had so much knowledge of the competition rules that he was invaluable as the event steward. It was his duty to adjudicate on any protests made by the competitors and his decisions were never challenged.

The Avro club was affiliated to an umbrella organization called the Association of Manchester Industrial Motor Clubs. Most of these clubs were quite small and very few were officially registered with the national motor sport body; being registered enabled their members to enter events open to other clubs. The Avro club (which despite the change to Hawker Siddeley Aviation kept the Avro name) was registered, which also allowed it to organize competitions open to members of other clubs. One local club that was not registered was Imperial Chemical Industries (ICI). It was a large organization and had several keen competitors looking for a suitable club to join. For some reason lost in time, the constitution of the Avro club did not restrict membership to the Woodford staff. Derek knew some of the ICI members and, keen to enhance his pool of competent organizers, invited them to join. Word soon got around and not only ICI people but some from the other Manchester clubs as well as a good collection of unattached car enthusiasts were applying to join. Very soon we had become a thriving competitive club. Some of the traditional Avro club members were uncomfortable with this situation and an attempt was made to change the constitution. At that time I was the chairman and I managed to put together a well-balanced committee, enabling both factions to live with the new status quo. The extra membership fees enabled more tools to be purchased.

I was becoming quite a successful rally navigator but, just like Derek, I was itching to try my skill in the driving seat. Although I had used my Singer Gazelle convertible with Derek for checking possible rally routes, it was not really much use as a rally car. The main reason was that the under-floor strengthening, necessary to compensate for the lack of a roof structure, made the ground clearance pitifully small; not good for competitive use on unmade roads. It had to go. I sold it and bought myself a second-hand Mini Cooper. This was only a partial success. We now had several competitive drivers in the club with powerful rally cars who were desperate for success and needed a good navigator. There were Ford Anglias fitted with 1,500cc engines, an Austin Healey 3,000, a Brabham-tuned Vauxhall Viva and a few members had acquired the Mini Cooper S or the new powerful Ford Escort. These opportunities for success as a navigator were too tempting. The standard Mini Cooper had a 997cc engine with two carburettors. It was more powerful than a normal Mini but, although it had disc brakes on the front, they were pitifully small and prone to fading when used hard. One of the things a rally car needed to have was good brakes.

I longed for a Cooper S but the cost, even of a second-hand one, was out of my league. However, I had a plan. I would strip all the parts out of the Mini Cooper, and have the rusting sections of the body shell replaced with some welding done to strengthen it. One of the newer members of the club had a father who ran a garage business with all the necessary equipment and he agreed to do the work for a reasonable sum. My old shell went away on his trailer, but my limited cash flow delayed its return for several months. In the meantime I had to have some transport, so I brought an old Mini Van which was the source of much amusement.

Eventually the body shell returned, painted in a gleaming red; I had concluded that the only appropriate colour for a Mini was red. In the meantime, I had been accumulating many of the special Cooper S parts ready for fitting. The front and rear sub-frames from the Cooper were in surprisingly good condition and, after a bit of strengthening, were the first parts to be fitted. A lot of special Cooper S suspension parts, the disc-brakes and drive shafts had to be newly-purchased.

This set my budget back somewhat. In order to raise the normal ground clearance slightly and change the camber angle of the wheel alignment for optimum handling, some small bushes needed to be fitted in the suspension linkages. I knew exactly what was required, and to save some money, one of my friends in the Woodford machine shop kindly made me a selection of different sizes to try out. The main expense would be the engine and gearbox which would have to wait. In the meantime, I fitted the 997cc Cooper engine to try it out.

Before I could use it on the road, I had to get it registered. Because it was classed as a rebuild, it was eligible to be newly-registered. Some form-filling, a new chassis/build number which I invented, a brief inspection by a Department of Vehicle Registration engineer and all that remained was to be allocated a registration plate number. For registration in Cheshire the last two of the first three letters were fixed and it had to end with a 'J' (for the year of 'manufacture'!), but I had some available choice for the rest. I chose XTU 900J. Some months later, when I had saved enough, I found a professionally rebuilt second-hand 1,310cc Cooper S engine and gearbox which completed the project.

The author's Mini Cooper S night rallying. (*Author*)

I did enjoy driving XTU 900J and had some success in competitions. The choice of suspension bushes resulted in an amazingly stable car, particularly on unsurfaced roads. However, the cost of doing a lot of competition driving was rather expensive, so I also continued with the navigation.

After the Andover test work was completed at Woodford, there was a lot of report-writing to do for the submissions to the MoD. I was happy producing graphs and charts, but my report-writing tended to be slow and laboured; a hang-up from the O-level English difficulties. Some days the words would flow quite well, but I particularly remember coming to the end of one day and being greatly embarrassed to realize I had barely completed a single sentence. Ken Wood was extremely patient and ultimately all the reports were signed off in time. In the middle of this, Squadron Leader Ogilvie, the RAF Andover project pilot, had visited Woodford to make a couple of assessment flights and had pronounced himself satisfied.

In the early part of 1967, there was little flight-testing activity. However, there were a series of modifications being incorporated onto the Shackleton Mk 3 as well as the old tail-wheel Mk 2. The MoD had a pressing need to introduce some of the latest surveillance and submarine detection equipment – ultimately to be installed in the Nimrod – into operational use. Utilizing these old Shackletons for this purpose was their only option at that time. Most such modifications needed flight-testing and I happily participated in some of this work.

The Nimrod was developed from the de Havilland Comet civil airliner and designed in response to meet the RAF requirement to replace the Shackletons. The original Comet's Avon turbo-jet engines were replaced with Rolls-Royce Spey turbo-fans, with more thrust and much better fuel efficiency, particularly at lower altitudes. The systems were basically unchanged from the Comet 4s. The main external change was the addition of a bomb bay and a large nose radome. Later a Magnetic Anomaly Detector (MAD) boom was added, extending from the aft fuselage. XV148, and the second prototype to fly, XV147, were built from two unfinished Comet 4C airframes. XV148 was aerodynamically representative of the production Nimrod, but XV147 retained the Comet's Avon engines and was used primarily for systems development. The first flight of the Nimrod prototype XV148 was on 23 May 1967 and was from de Havilland's assembly airfield at Chester to Woodford. It was flown by the de Havilland chief test pilot, John Cunningham. Jimmy Harrison, Avro's chief test pilot at the time, would have normally been expected to fly it but John Cunningham had done the flight-testing of the Comet 4. Jimmy was the co-pilot. Martin Garland was the flight test observer on this and most of the flights of XV148, although Mike Turner also did

Hawker Siddeley Nimrod prototype XV148. Photo taken before fitting the enlarged strake ahead of fin to improve directional stability. (*Avro Heritage Museum*)

quite a lot of this work. The versatile Bob Pogson (Avro's AEO on Vulcan) had retrained to be the regular Nimrod flight engineer.

From the wind-tunnel tests, it had been anticipated that the existing Comet fin would not be large enough to balance the additional fuselage side profile of the bomb bay. A few weeks after the first flight, it was time to fully explore this potential problem. The design engineers anticipated that there could be a point where the fin was likely to stall and if this happened, the aeroplane would lose all directional stability; the aeroplane could enter a potentially catastrophic uncontrollable spin. However, to gain sufficient information to determine how much bigger the fin needed to be, this testing had to explore this situation as far as was reasonably safe. Mike Turner was the flight test observer assigned to this flight on XV148, but he wanted to be able to concentrate on plotting the test points as the rudder was steadily applied to increase the sideslip angle. I was drafted in to work the instrumentation console, so Mike was free to occupy the spare cockpit seat with his graph paper. As usual for such stability tests, the aeroplane was loaded close to the aft c.g. limit. Because of the potential danger, all the crew wore parachutes. I had a rudder position indicator, a sideslip indicator and a rudder force read-out on the console. A temporary sideslip indicator was also fitted onto the flight-deck instrument panel.

Jimmy Harrison was in charge. When he was ready to begin, I started to run the recorders and he moved the rudder pedal forward a few degrees and held it steady. I read out the rudder angle and rudder foot force, while Mike plotted these against the sideslip angle. This was repeated with three or four more gradual increases in rudder deflection. From Mike's plot, it was noticeable that the sideslip was beginning to increase proportionally more with the last rudder increment. The curve was still stable, but caution was required in going any further. Jimmy elected to try a further small push on the pedal. As soon as I had read out the rudder readings, without hesitation Jimmy stopped the test and returned the rudder to neutral. Mike's graph clearly showed the sideslip had increased much more and everyone agreed that we had gone far enough. To go any further would have extended the academic comparison of actual and predicted results, but safety dictated otherwise. We repeated the test with the rudder in the opposite direction with almost identical results; we did not feel the need to go quite as far. Job done.

From the detailed analysis of the instrumentation results, it was determined that fitting a large strake in front of the lower part of the fin would be sufficient. There had been concern that a totally new fin might have been necessary.

I did a little bit more Nimrod flying but my next project was the Series 2A 748: an increased power version of the previous 748 models. The first prototype Series 2A, G-AVRR, flew in 1967 and soon after it became almost my only flight test project for the next six months. The 748 had several stability and control features that only marginally met the ARB requirements (BCAR). Increasing the power was likely to make some of these issues more critical. It was a demanding programme and Tony Blackman was again the project pilot. The rate of flying was intense, and it proved to be a very reliable aeroplane. In the six months I did approximately 140 flying hours, interspersed by a handful of Nimrod and production 748 flight tests. It was certainly a challenging and rewarding programme.

The real icing on my cake was the tropical trials. All civil aeroplanes are offered to airlines with the capability to operate in high-temperature conditions and from high-altitude airfields. This necessitates undertaking actual testing in representative conditions. In such conditions, it is the take-off distance and rate of climb that are most critical and must be measured; normally, the domain of the Performance Section flight test staff. Some months before these trials on the 748 2A were scheduled to commence, Bill Horsley had left Avro (HSA) to join Andrew

McClymont at the ARB and no other performance observers were available. In this situation, Mike Turner recommended that I should join the team to manage the instrumentation console while he would mastermind the programme.

The location chosen for these trials was Ethiopia. It was hot enough even in January and the main airport at Asmara was at a height of 7,600ft and close to the equator. There was a second advantage: we were also able to operate in very high temperatures at a sea-level airfield on the Red Sea coast at Massawa, not far away.

On a cold January morning in 1968, we set off early for our first refuelling stop at Rome's Ciampino Airport. I walked down the steps from the aeroplane onto the concrete.

I said to myself: 'I am 26 years old and this is the first time I have stepped onto foreign soil; even though it is concrete!' I didn't share this thought with anybody else at the time.

Having refuelled, we set off for a night stop in Nicosia at the Cyprus Hilton. There were ten of us in the team. The project pilot was Harry Fisher. He had with him as co-pilot Steven Cherry-Downes; relatively new to the Avro staff and recruited primarily as a communications pilot to fly the company de Havilland Dove. (This aeroplane was frequently used to transport staff between various airfields.) Bob Dixon-Stubbs was the navigator and 'Mr fix it'. He had planned the route, organized all the hotels and carried a large quantity of cash, in various denominations, to hand out as necessary and pay the bills. (Credit cards were not yet in common use outside the USA. However, the major suppliers of aviation fuel, such as Shell, issued their own cards, known as 'Carnets', to reputable airlines and aviation companies which were used to purchase fuel.) Mike Turner and I would be flying on the tests. Mike Taylor, head of the Flight Test Performance Section, with one of his staff would do as much analysis of the test results as was immediately necessary. He was also in charge of the ground camera, which was used to record the take-offs and landings. To assist him, he had a member of the photographic section who would operate the camera and process the films. There were two Avro ground engineers to look after the aeroplane and help with many of the other tasks. To make sure the engines were operating up to specification, we also had Gordon Fox, the Rolls-Royce representative who joined us in Asmara. In this mixed and very amiable group, I settled into a new life of 'posh' hotel living without much difficulty.

The next morning we took off en route for a refuelling stop at Cairo. Cairo was certainly an experience to remember. We were marshalled into a remote part of the airfield where there was a scattering of refuelling points set in the tarmac. The pilots and Bob Stubbs disappeared off to do the paperwork and pay the bills. The rest of us were left to watch the proceedings of the airport ground refuelling chaps. It was soon obvious to us that none of the fuel hoses that they had were going to reach the aeroplane from any of the ground fuel points. Our ground engineers tried to help but without the pilots, we could not move the aeroplane. It appeared that there was no means of coupling any of the hoses together. After some time spent trying different hoses in different locations, a small fuel bowser arrived. They positioned it in between one of the fuel points and the aeroplane, connected one hose to the bowser and another one from the bowser to the aeroplane. Eventually this worked, once they realized that they had to partially fill the bowser before it could pump fuel to the aeroplane. When the pilots returned, they could not understand why we were not fuelled up and ready to go. We had a few laughs recounting the saga.

Avro 748 Series 2A G-AVRR at Asmara for tropical trials, 1968. (*Author*)

Tropical trials team outside
Asmara hotel, 1968. L to R:
Steven Cherry-Downes, Mike
Taylor, the author, unknown,
Harry Fisher, Mike Turner,
unknown. (*Gordon Fox*)

Asmara was an extremely pleasant city. Everyone was so friendly and helpful. It seemed to be a perfectly safe place in which to wander around. Bob had no difficulty in laying on transport for us to and from the airport every day, and the airfield staff did everything they could to help with setting up our cameras. It's so sad to think of what happened to Ethiopia years later and know that it will probably never be the same again.

We had to use sandbags to ballast the aeroplane to maximum weight for the tests. The amount of help the airport staff offered to fill these bags was almost embarrassing. We carried our own set of scales for measuring the weight of each bag and always made sure the total was correct. The bags were strapped down in the correct location to keep the c.g. near the critical forward limit for performance testing. On our deployment to Massawa we carried as many ready-filled bags as we required, but we still got all the help we needed to move them about.

The days were filled with measured take-offs, landings, accelerate-stops and rate of climb measurements. All the runway tests were recorded by the ground camera which tracked the position of the aeroplane. The camera was calibrated with fixed points on the ground, so that the position of the aeroplane at any time could be determined from the film record. The take-offs and accelerate-stops had to be performed close to the maximum take-off weight, and the landings close to the maximum landing weight. For structural reasons, it is usual for the maximum landing weight to be quite a bit less than the maximum take-off weight. It is not necessary to refuel back to the maximum weight before each take-off as the measurements over a small spread of weights can be extrapolated to determine the final figures. However, after each take-off, a landing is necessary before the next take-off measurement is made. Measured landings are not made above the maximum landing weight, but when test-flying and where the difference between the maximum landing and take-off weights is not great, the test pilot is authorized to land above the maximum weight with due care in order to be back on the runway ready for another measured take-off.

For the take-offs, the engines are run up to full power and then on a countdown, the brakes are released. At the decision speed known as V_1, one engine is failed (shut down) and the take-off is continued. At the rotation speed (V_R), the nose is eased back to reach the take-off safety speed known as V_2, at an altitude of 35ft. V_1 is the speed below which, if an engine fails, there will not be adequate controllability or sufficient remaining runway to safely continue the take-off. If an engine fails at or below V_1, the take-off has to be abandoned with the brakes and reverse thrust rapidly applied to bring the aeroplane to a stop before the end of the runway. Once V_1 is reached the take-off must be continued, even if an engine fails immediately after V_1. To provide the necessary margins for safety, these speeds are limited as follows:

- V_1 must not be less than the minimum control speed on the ground (V_{MCG}; see the explanation in Chapter 4). The airline pilot may not be expected to achieve the maximum lateral deviation of 30ft as required in the establishment of V_{MCG}, but the width of commercial runways provides a significant additional margin.
- V_R must not be less than V_1 nor less than a speed 5 per cent above the measured stall speed in the appropriate take-off configuration; it would not be good to stall during the take-off.
- V_2 must not be less than 20 per cent above the stall speed, nor less than 10 per cent greater than the minimum control speed in the air (V_{MCA}; again see Chapter 4), nor less than 8 per cent above the stall warning speed in the take-off configuration.

The measured landings were a bit more challenging. The official landing distance starts when the aeroplane is 50ft (15m) above the runway surface. We had a radio-altimeter on the pilot's instrument panel; radio altitude was also recorded on the instrumentation. It was also determined from the ground cameras used for recording the landing. The target airspeed at the 50ft point is known as V_{AT}.

This speed varies if the approach is made with an engine inoperative. (In the case of aeroplanes with more than two engines, the case with two engines inoperative also has to be considered.) With all engines operating, this speed (V_{AT0}) may not be less than 22 knots above the stalling speed in the landing configuration, nor less than the minimum control speed in the landing configuration (V_{MCL}; again see Chapter 4), nor less than 8 per cent above the stall speed in the landing configuration. With one engine inoperative, this speed (V_{AT1}) may not be less than V_{AT0}, or less than V_{MCL} plus 5 knots.

Once below the 50ft point, the pilot has to place the aeroplane firmly onto the runway (but not so hard that it bounces back into the air), step hard on the brakes and select reverse thrust. Approaches are made with both engines operating and with one engine inoperative. The tests with only one engine in reverse thrust will give a less favourable landing distance result, and the ability to maintain directional control had also to be assessed.

The accelerate-stop and landing distances are very punishing on the brakes, so even though the numbers of these tests are kept to a minimum, the spare brake pads carried with us needed to be fitted regularly.

In theory, the measurement of rate of climb performance is more straightforward. The aeroplane is set up in the appropriate configuration with one engine inoperative and the other at maximum power. Once stabilized, the climb is continued for at least five minutes to measure the height gained. The engine power is affected by atmospheric conditions such as altitude and temperature. This is the reason why the performance needs to be measured in hot and high conditions. In normal atmospheric conditions, the air temperature reduces at about 2 degrees C per 1,000ft. A five-minute climb will cover several thousand feet and for a meaningful result the actual temperature needs to be reducing close to this rate. However, particularly in tropical conditions, it is not uncommon to find sections of atmosphere where the air temperature may actually increase with altitude. This is not good for climb-testing because the actual engine power will not be varying as expected during the climb and, generally, the achieved rate of climb will be significantly less than it should be. This means that a lot of time may be spent looking for atmospheric conditions close to the accepted norm.

It was normal for the ARB to send one of their performance engineers to monitor the company's tropical trials, to check the testing methodology and sample the results. Bill Horsley had only just joined the ARB and Andrew McClymont was involved in other overseas work. The only other performance man they had available was Mike Smith. Mike was fundamentally a helicopter expert and possibly less than an ideal choice but, after his arrival at Asmara, it was not long before his competence was never in doubt. He was a very cheery, slightly overweight, extrovert character who was the source of much entertainment, but he asked enough searching technical questions to get us thinking. He took the measured distance results, plotted them on his own graph paper and drew his own 'best fit line' through the points. If this differed from ours, then a compromise had to be agreed.

One evening towards the end of our stay, we all went out to a restaurant some distance from the hotel. We used taxis to take us, but Mike Smith had a better plan for returning. The Ethiopian alternatives to taxis were horse-drawn 'chariots' called 'gharris'. They could carry two passengers plus the driver. Mike, with the enthusiastic support of Gordon Fox, unbeknown to the rest of us, had organized four of these to arrive at the restaurant at the time we wished to leave, and he encouraged the drivers to race back to the hotel. On the cobbled streets, with the horses' hooves and hard wheels sliding about, it was undoubtedly the most dangerous part of our visit. Surprisingly no-one fell off and we did all make it back. It was not a great surprise that Mike Smith, accompanied by Gordon Fox, chose the winning gharri. The description of this event written by Gordon Fox in John McDaniel's book[1] adds an extra aspect to this story of which I was not personally aware and could not comment!

After ten days and some sixteen hours of flying, we set off back to Woodford via Cairo, followed by a night stop in Malta and a refuelling stop in Nice.

There was still some stability and control testing to be done and then a visit from the ARB to go over our results and select some aspects they would like to flight-check themselves. Unlike their previous visit on the Andover, this was a critical milestone in the certification process. If they found anything they deemed to be unacceptable, then design changes and/or unfavourable revisions to scheduled operating speeds may have been necessary; this would be a major setback to the programme. The ARB team this time was pilot John Carrodus with Keith Perrin. Tony Blackman occupied the right-hand (co-pilot) seat. All went well until their last flight which terminated with the inevitable assessment of the minimum control speeds. The airborne checks of V_{MCA} and V_{MCL} went OK but when it came to check the V_{MCG} on the runway things became difficult.

Asmara gharri transport, 1968! (*Gordon Fox*)

During the take-off run, Tony failed the critical engine at our scheduled V_{MCG}, whereupon it was then up to John to control the lateral deviation and continue with the take-off. On John's first attempt, the lateral deviation was estimated to be about 70ft, well in excess of the 30ft requirement. Even without consulting the instrumentation records, it was possible to make a fairly accurate assessment using lines marked on the runway. Before the second attempt, Tony explained to John that he had found it necessary to use aileron as well as rudder for the optimum directional control of the asymmetric thrust. John had another shot at it. It was better, but the deviation was still about 50ft. A further discussion followed and, very tentatively, Tony offered to demonstrate his technique. I think John and Keith were a little taken aback, but it seemed to be the only sensible way forward. Tony and John swapped seats and with John fully instructed on how to rapidly shut down an engine, off they went. At the exact speed, the critical (right) engine was failed; Tony's arms and feet were, as later described by John, 'a blur'. The right main wheel temporarily left contact with the ground. The port wingtip, on the side with the operating engine, got quite close to the ground. The lateral deviation was estimated to be about 15ft; comfortably less than 30ft.

Tony said: 'Well, that's the technique; I think you should have another go.' John was not going to disagree. They swapped back seats and at the engine failure, John applied full rudder and as much aileron as he dared; clearly not quite as much as Tony had done, but the deviation was contained to about 25ft.

Before the post-flight debrief, John and Keith felt they needed a little time to themselves to think over the events. We waited with much trepidation. In the debrief they went over all the results of their flight assessments, pointing out those marginal items that they would require us to include in the Production Flight Test Schedule. Then they came to the V_{MCG} assessments. John's view was very salutary:

> If it is possible to keep the deviation within the 30 feet, then it has to be concluded that the requirement had been met. Most pilots are not 'Tony Blackmans', but the 30 feet was set deliberately low to ensure that less experienced or less capable pilots should still be able to keep the aeroplane within the width of the runway if faced with this very demanding situation.

Many years later, I came to know John Carrodus very well. He was undoubtedly one of the best engineering test pilots. Sadly, he is no longer with us.

Note

1. McDaniel, *Tales of the Cheshire Planes* (1998), pp.91-92.

Chapter 6

An Unusual Measure of Success

The first production HS 748 Series 2A, G-AVZD, flew in March 1968. A lot of flying was necessary to confirm its handling characteristics were representative of G-AVRR and complete all the certification testing. There were no new difficult areas and, true to their word, the ARB came along to re-check the critical points. Again, it was John Carrodus and Keith Perrin. The V_{MCG} test was on their list, but John was ready this time. A rapid application of full rudder and a healthy degree of aileron kept the deviation in check. Honour was satisfied and certification granted.

By this time 748 Series 2As and Andovers were rolling off the production lines. A lot of production flight-testing had to be done. In addition, there was the frequent introduction of small modifications to the production standard, many of which necessitated some specific flight assessment. However, we were beginning to find that some of the Series 2As were having difficulty meeting the critical stability cases and the stick force per g requirement test. This was a big headache. Without being able to sign off the Production Flight Test Schedule, a Certificate of Airworthiness would not be granted. An explanation had to be found. By a fortunate coincidence I was in the hangar when one of the hangar ground engineers pointed out some variability he had noticed in the way the elevator trailing edges had been fabricated. In some cases, the aluminium skin between the ribs bowed upwards quite noticeably, while on others it was fairly flat.

We in the Flight Office were grateful for any possible reason that might explain these handling problems, so we set to work with some calculation as well as a search through relevant published data. Someone found an old report in the Royal Aeronautical Society Archives, which showed that the aerodynamic lift of the elevator was affected by the trailing edge angle of the skin. Eureka! We manually pinched together the trailing edge skins on the elevators of one of the problem aeroplanes and set off to re-check it in flight. It passed the tests. We had a template made which when fitted over the trailing edge would show how much 'pinching' was necessary. In some cases, we overdid it and the opposite treatment was necessary; we obtained a suction device with which any overly concave profiles could be sucked out. This problem was reported to the Production Quality Control people but the next batch of elevators to be manufactured was some way off. I was not aware at what stage the Production Department took control of this issue and what happened to those elevators already in store. The next few production aeroplanes still had some problems, but in the flight-test hangar we had this fully under control and no aeroplane would get its Certificate of Airworthiness and be delivered without being suitably 'fixed'. I did wonder about what would happen if an aeroplane in service damaged its elevator and had to be fitted with a new one that had not been subjected to the appropriate check. However, we were talking about small degrees of non-compliance coupled with a string of circumstantial possibilities not too likely to happen, so no-one seemed to be too worried.

Around that time, one of the senior members of the Handling Section, Geoff Roberts, left to join the Atomic Energy Authority in Oxford. He was not seeing much opportunity for advancement at Avro and decided he needed a change of direction. This worked out very well for him. The amount of workload at Woodford meant he needed to be replaced and Geoff Stilgoe joined the section; like me he had done the university/apprentice 'sandwich course'. He soon became a valuable addition to the section and, also like me, he joined the band of flight test observers.

My next major flight-test project was the installation of the latest Smiths Industries autopilot, the SEP 6, into the 748. The Civil Aviation Flying Unit (CAFU) was responsible for the calibration and commissioning of the localizer and glideslope landing approach guidance beams at UK civil airports: the Instrument Landing System (ILS). Some UK airports were installing the newer, more accurate ILS equipment, necessary for the automatic landing system being developed on the British Airways Trident aeroplanes. In order to be able to accurately assess these new ILS beams, CAFU needed an aeroplane and autopilot combination with the appropriate capability. The 748 was of a suitable size and the new Smiths Industries SEP 6 autopilot was state-of-the-art. CAFU placed an order for two aeroplanes. At the same time the German government equivalent organization to CAFU, the *Deutsche Flugsicherung* (DFS), had the same requirement and also chose this 748 variant.

The SEP 6 autopilot came with an integrated SFS 6 flight director system. This is a system that does all the computations of an autopilot, but instead of sending its output to the aeroplane control servos, it presents the required demands onto an instrument known as a 'flight director' which the pilot can follow manually. For convenience, this flight director is incorporated into the normal artificial horizon attitude display. The benefit of this is that the pilot is able to remain in direct control of the aeroplane while following the commands computed by the system to maintain the required flight path. For example, on an instrument approach to land, where the pilot is controlling the aeroplane manually, by following the flight director commands, at the point where he is committed to complete the landing, he is already flying the aeroplane and is not faced with the sudden requirement to take over manual control as he disengages the autopilot. It can also provide useful and helpful guidance immediately after take-off, where the autopilot cannot be engaged before a steady climb has been established.

This flight-director concept was conceived some years earlier but was not without problems. When manually flying, in order to achieve the required flight path, pilots had become practised at interpreting the available information from their instruments, such as altitude, airspeed, attitudes and deviations from the radio beams. Presenting the pilot with a display that would 'direct' their actions introduced a different concept. For instance, if the flight system decided that the pilot had to apply some bank angle to achieve a new heading or correct a heading error, it ideally needed to show him the bank angle to achieve. Many years before, an early flight-director system was shown to the ARB Chief Test Pilot Dave Davies for his assessment. This story was quite well-known at the time and often repeated within the industry. Initially, he was happily following the command bar in roll and pitch to achieve the 'directed' flight path. After a short pause, they set up another situation. Dave was now more relaxed and as soon as a new roll command was displayed, his reflexes kicked in and he immediately interpreted the horizon attitude as an 'indication' and rolled the aeroplane in the direction to correct this perceived bank error; the opposite way to which the 'director' was intending. That was the end of that way of presenting the flight-director information. If a pilot as experienced as Dave could make that mistake, inevitably many pilots would do the same.

This problem of knowing whether a display is giving you an 'indication' or a 'direction' is quite common in some everyday situations. A good example is the modern single-lever wash-basin taps. You look at the tap and you are usually confronted with a red and a blue mark. If you want hot water, do you take the colours to be 'directing' you to move the tap towards the red mark (to the left), or is it acting as an 'indicator' telling you to move it in the opposite direction to show you the red mark? By international plumbing convention, most of us have learned that hot is to the left and cold to the right. You need to move the tap in the direction of the appropriate colour: it is a 'director', but how would a visitor from another planet know this?

Collins, an autopilot manufacturer in the USA, devised a particularly successful solution. They made the director section of the display into a 'V-bar' shape. It moved in pitch and roll in response to the demands of the system. Below this was a symbol, clearly representative of a delta aeroplane. The pilot has to continually 'fly' this symbol to fit into the V-bar; simple, unambiguous and intuitive. They had the sense to patent it and Smiths had to find another solution. The SFS 6 did have a little aeroplane symbol, but it had to be 'flown' to follow a pair of moving crossbars; one vertical and one horizontal. It worked, but the Collins solution was the worldwide preference and the SEP/SFS 6 inevitably had many fewer customers.

Flight-testing the first CAFU 748 Series 2A, G-AVXI, commenced in February 1969. Initially we had to complete the normal Production Test Schedule checks before getting stuck into the autopilot work. One of the new test pilots, Stuart Grieve, did the routine flying but Tony Blackman was not going to delegate the SEP 6 development test work. Tony was soon to become chief test pilot on Jimmy Harrison's retirement from flying. Jimmy had decided it was time to move on and became the product support manager, a post he fulfilled with enthusiasm and great success. As with our previous SEP 2 work, we took the lead with Smiths Industries providing close support. The test-box was more complicated, with many more possibilities for our adjustment. Setting up the autopilot for normal en route situations was the first task. Smiths had recommended some initial settings for the control gearings and time delays, and these were duly set into the test-box. Being a new autopilot with more scope for adjustment, we took quite a long time to reach our considered optimum settings. Then there were the failure runaways and oscillatories to check. Although the SEP 6 was an advanced autopilot, it was still single-channel, unlike the three-channel systems being developed for automatic landing. The SEP 6 had some improved internal monitoring over the SEP 2, but this could not prevent the critical failures. All this test work took quite a long time to complete. The next and more interesting testing was to determine the optimum autopilot computer settings to ensure that the localizer and glideslope beam tracking would achieve the necessary high accuracy when following the airport's Instrument Landing System (ILS).

The International Civil Aviation Organization (ICAO) had a set of weather criteria categories for approach and landing. There were three categories. Category 1 required an aircraft to have the capability to operate down to a minimum Decision Height (DH) of 200ft with a minimum Runway Visual Range (RVR) of 800m (2,624ft). This was the best that could be achieved by the earlier state–of–the–art autopilots. The aeroplane had to arrive at the 200ft point in a position and attitude from which the pilot could visually continue to a safe landing; the acceptability criteria for this were set in the requirements. Category 2 extends the DH down to

100ft and the RVR to 400m (1,312ft). Category 3a is for automatic landing with zero DH and an RVR of 200m. Category 3b is zero DH and 50m RVR.

It was decided that the minimum approach beam tracking criteria for the calibration aeroplane would be to achieve this Category 2 accuracy. Most of the necessary flight-testing could only be done using an actual airport ILS. At that time, there were very few airports that had installed the latest standard: Heathrow, Manchester, Chester (Hawarden) and Hatfield; the latter being the Hawker Siddeley airfield where the Trident auto-land was being developed. Heathrow and Manchester were usually too busy to accept a non-scheduled aeroplane doing repeated approaches and so we used Chester mostly, it being conveniently close to Woodford. We had to be sure that we were not unreasonably 'tuning' our system to one beam, so we used Hatfield and, briefly, Heathrow and Manchester as well. Although the 748, being rather twitchy and responsive to turbulence, was not the most ideal aeroplane for this task, we managed to achieve total success. As with the en route autopilot work, the failure cases also had to be tested. In the case of an aeroplane flying a 'coupled approach' to land, the pilot was considered to be more attentive; consequently, the response delay for failures was reduced from two seconds to one. The critical approach case was the nose-down runaway. The amount of height lost during this failure and recovery had to be considered in the determination of compliance with the 50ft DH criteria. Without having to consider possible failures, the autopilot was easily capable of controlling the aeroplane down to less than 50ft.

During this autopilot work, we were also setting up the SFS 6 flight-director control laws. The pilot's recognition of the errors requiring correction was not always the same as those of the autopilot. However, when presented with a director display that showed him how much pitch or bank was required to acquire and follow the required flight path, his responses were quite like those of the autopilot. We demonstrated that we could also achieve the Category 2 criteria with the flight director, a result of which we were particularly proud.

This was new certification work and the ARB made their customary assessment visit; it was John Carrodus again and he was very impressed. The CAFU pilots also came to make their own checks, and they too were very pleased. Soon afterwards, the first German aeroplane D–AFSD also joined the programme and it contributed significantly to the amount of approach flying on the Category 2 beams. This whole SEP/SFS 6 programme took about ten months to complete.

Just in case I might have been getting bored during the 748 programme, some Shackleton test work was going on and I was called upon to participate in this on a few occasions. One of the new systems that was to be installed in the Nimrod was a

Avro Shackleton Mk 2 WL754 with forward radome. (*Avro Heritage Museum*)

very advanced search radar system. The MoD was again desperate to introduce this into service as soon as possible, but the operational Nimrods were still years away. Again, the trusty old Shackleton was called upon to fill the gap. Now, however, all the Mk 3s were fully committed and some had been grounded because their flying hours were stretched beyond the fatigue life limit. That left only a few of the tail-wheel Mk 2s that could be made available. Work went ahead rapidly to fit the large radome into the nose of WL754 and testing commenced. Bill Else was again the project test pilot and, although I was needed to fill in for the other flight test observers on only three flights, the work was certainly different.

Over the next two and a half years, and immediately after the termination of the 748 SEP 6 programme in December 1969, I began a long and varied involvement with the ongoing Nimrod test programme. Amazingly, it had been about two years since my previous test flights in the Nimrod when we had sorted out the directional stability problem by increasing the fin area. The second of the prototypes constructed from the unused Comets, XV147, was still in regular use for ongoing test work, particularly systems testing. However, since mid-1968 the production Nimrods had been flowing slowly off the assembly line; the integrated anti-submarine systems needed much testing and development work.

At the heart of these systems, each Nimrod had one of the first commercially-available digital computers. It was manufactured by an American company called 'Digital'. This computer was not only used to integrate and control the on-board systems, but it acted as a recorder of all the data involved. This it transcribed onto magnetic tape. 'Digital' also provided one of these computers for use on the ground, primarily to analyse the information from the on-board magnetic tape recordings. Derek Bentley oversaw this analysis; it did not take him long to appreciate the value of such an asset for the analysis of other flight-test work. With the guarantee that none of this 'moonlighting' work would interfere with the Nimrod programme, he got his wish to make use of it. We had a new recruit in the Flight Test Section, Bob Ollerenshaw, who was very good with computers and Derek, with his help, developed a series of programmes so that the 'Digital' computer could analyse some of the non-Nimrod testing, in particular 748 Category 2 results. Not wishing to be left behind, I learned some useful programming skills using the Fortran computer language, then very new. This proved to be an extremely useful attribute.

I did a bit of flying in XV147, but the second production aeroplane XV231 was ready, closely followed by XV232. XV226 was actually the first production Nimrod. It first flew in June 1968, but it had not been fully configured with operational equipment and so was being used for an assortment of development testing. Ted Hartley and Martin Garland were the flight test observers primarily responsible for this work, so I did not get to fly in it until much later. As mentioned earlier, Bob Pogson (Avro's AEO on the Vulcan) had re-trained to be the regular Nimrod flight engineer. When I first went on board XV231, one of the first things I noticed was how few windows there were. The two prototypes still had all the windows from the Comet, although many were invisible behind equipment. In the production aeroplanes, the flight test observer's console was sideways-facing with no window in sight. In the operational configuration, most of the flight crew stations also faced sideways and for normal flying it was no problem. However, carrying out flight-test manoeuvres and stall tests in this position was noticeably more nausea-inducing. Fortunately I was OK, but some of the other flight test observers were reluctant to go on any flight where rapid manoeuvring was on the programme. Consequently, I did a large percentage of such flying.

The main focus of flight testing at this stage was to establish the accuracy of the self-contained on-board navigation system. For this purpose, the Nimrod was

HS Nimrod XV226, first production aircraft used for development and certification, with the enlarged strake in front of the fin and MAD boom extending from the aft fuselage. (*Avro Heritage Museum*)

fitted with an Inertial Navigation System (INS). It was many years before the advent of satellite navigation systems. At the heart of an INS were three rapidly-rotating gyros in axes aligned with the fore and aft, the vertical and the lateral directions. These gyros would react to any small acceleration in any direction and each such reaction was detected and analysed to compute the movement of the gyros from their starting-point. There were two particular difficulties with the INS systems. Firstly, the gyros had to be aligned in a known direction (heading) and, secondly, such gyros were prone to drifting slowly from their initial alignment over time. It would be a demanding goal to maintain accuracy over the many hours' duration required of the Nimrod.

The alignment issue was solved by using the take-off runway heading as the starting datum. The accurate heading of the runway was set into the navigation system computer and the pilot did his best to track along the runway centreline during take-off while the computer registered this initial heading as the datum. Whenever a flight was specifically made for the purpose of checking navigation accuracy, any error made by the pilot in following the centreline was measured by firing a paint blob from the aeroplane onto the runway at the point where the navigation computer completed its alignment. Ground staff measured the distance of the aeroplane from the centreline before take-off and, after take-off, the distance of the paint blob. From this, the true heading tracked by the Nimrod could be calculated accurately. The effect on the subsequent navigation accuracy of an average, typically very small, piloting error in maintaining the actual runway heading could be statistically evaluated. Determining the effect of the gyro drift was to consume very many flying hours.

The operational concept of the Nimrod was seen, by the MoD and RAF Coastal Command, as rather different from other aircraft. Its primary role was the detection and possibly destruction of enemy submarines. This required the management and integration of inputs from several on-board operators: navigator, radar operator, underwater sonar-buoys (dropped from the aeroplane for detecting underwater sounds), MAD and ultimately depth-charge deployment. In addition, the pilot had to be able to position the aeroplane in the correct location for these tasks. A normal crew complement was about twelve personnel, but up to twenty-five could be accommodated in exceptional circumstances. Somewhat controversially at the time, it was decided that the overall command of the crew should rest with the 'tactical navigator' (Tac Nav) rather than first pilot. The pilot would retain the responsibility for all flight safety issues. In reality, this was a good decision; it was the best way to successfully manage the complex missions.

Immediately aft of the flight compartment and crew entrance door, on the starboard side, there was a forward-facing console for two crew. The left position was for the 'routine navigator' who was responsible for the operation of the INS and all the en route navigation. To his right sat the Tac Nav; one of the few aft crew who had a window. On the console in front of him was a large circular TV (cathode ray tube) display. This could show the location of any sonar-buoys as well as any detected signals from them, or from the other radar and MAD search systems. Superimposed on this display were the aeroplane position and its recent track. Via the flight-director system on the pilot's instruments, the Tac Nav had the ability to directly input the control demands, which were computed by the on-board system, to direct the aeroplane flight path to best position it for the dropping of sonar-buoys or depth-charges. Obviously, the pilot had the ability to stop following the flight director at any time if he saw any safety issue.

At the outset, all the production Nimrods after XV226 were fitted with the complete navigation system as well as the nucleus of all the anti-submarine systems and search-and-rescue equipment. Therefore, to be able to operate these systems,

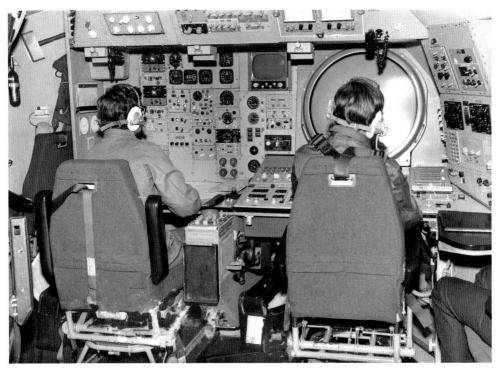

HS Nimrod navigation station, 1970. The author on left in routine navigator seat and Peter Morgan at tactical navigator station. (*Avro Heritage Museum*)

a flight could not be conducted without, at least, the routine navigator and tactical navigator stations being manned. Hence, in addition to any normal flight test observers for specific flight-testing work, extra crew had to be found from within the Flight Test Department to carry out these new flying tasks. Peter Morgan from the Flight Systems Section was chosen to be the lead routine navigator, primarily responsible for the INS and other navigation equipment, and I was asked to take on the Tac Nav responsibilities, although we were soon both able to perform each task. Sensibly there were back-ups also appointed. On many flights, it was possible to combine the Tac Nav task with other flight test observer duties, usually by changing seats to the instrumentation console and back as necessary.

On some occasions when positioning an aeroplane to or from an RAF base without any test commitments, one person could handle both routine and tactical navigation functions with a bit of seat-swapping. Both Peter and I were called upon to do this on a few occasions. I recall one such instance with Tony Blackman piloting, when we were bringing a Nimrod back to Woodford from the RAF Nimrod base at St Mawgan. It was snowing quite hard and Tony was itching to get into the air. I was the only one on board apart from the flight-deck crew. I had set the true runway heading into the system, gone through the Tac Nav pre-flight checks and moved across to the routine navigator seat, but the INS gyros were taking longer than usual to complete their self-checking process.

Tony was impatient: 'It's snowing harder now. We have got to go if we don't want to be stuck here. Let's forget the INS.'

'Give it a few more seconds; it must be nearly there,' I replied bravely.

I was always conscious of trying to make the most of any flying time, and every flight with this navigation system operating would be used in the statistical computation of the overall system accuracy.

Luckily for me, at that second the INS came to life. I checked it had integrated itself into the navigation system and said: 'OK, it's working; let's go.'

Tony muttered something in a grumpy voice as he released the brakes and applied full power. We comfortably made it into the air, but I was not sure whether my 'brownie points' went up or down that day.

Checking the accuracy of a navigation system was a very time-consuming process. The number of flying hours necessary to generate useful information was huge compared to the rate of data obtained during handling flight tests. There were two aspects to be covered by the testing. Firstly, the accuracy to be achieved after a long positioning flight to the search zone and secondly, the deterioration in accuracy when undertaking repetitive search manoeuvres. When operating in or close to the UK or other countries, which have systems to provide navigational information such as aviation radio beacons and the 'Decca-Navigator' system, the aeroplane's position can be determined to a reasonable accuracy. At that time, the Decca system was used extensively by military and civil aircraft. The Decca-Navigator system was developed towards the end of the Second World War and made fully operational in the UK in the early 1950s. A master transmitter, located just north of London, was surrounded by three equally-spaced 'slave' transmitters at a distance of about 80 miles from the master. Each 'slave' was allocated a colour for reference: red, green and purple. The beams generated by the master transmitter and each slave produced a pattern of radio waves. The Decca receiver in the aircraft had three instruments: red, green and purple to measure the beam signal. The read-outs from these instruments were plotted onto a map on which were superimposed the Decca lines to give an accurate position of the aircraft. While in range of such equipment, it was possible to determine and, if necessary, update any accumulated INS errors.

For the systematic assessment of en route accuracy, this Decca system was utilized. Fortuitously, the bisector of the purple slave beacon and the master

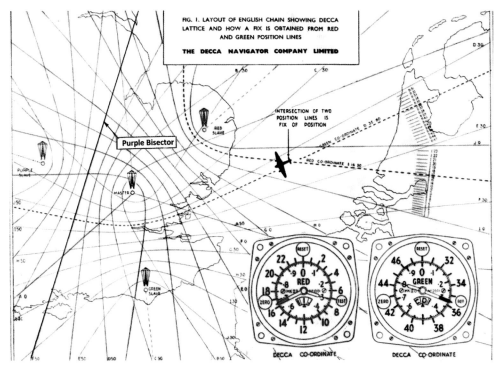

Decca navigator chart showing 'purple bisector'. (*Decca Navigator*)

crossed a large part of the UK from near Grimsby to Southampton. Known as the 'Purple Bisector', it was a straight line that could be followed accurately by keeping the purple indication constant while regularly reading/recording the red and green indications to compute the Nimrod's exact position. At the same time, the Nimrod's position computed by the INS was recorded, enabling a continuous direct comparison. We spent many monotonous flying hours tracking up and down this Grimsby to Southampton route; often the flight time was more than four hours.

However, when well out to sea, the Nimrod had to rely on its on-board INS navigation capability. Additionally, in a potential wartime scenario, the ground-based systems would be of use to the enemy and would, almost certainly, be turned off. To assess the INS system's ability to retain its accuracy during search manoeuvring, we had to devise a set of 'standard' manoeuvres typical of normal search patterns. These were 'circles', 'racetrack' and 'clover-leaf' patterns. The circles were continuous 15 degree banked turns. The 'racetracks' were straight legs of two minutes followed by a 30 degree banked turn through 180 degrees onto a parallel heading for a further two minutes with a similar turn in the same direction back onto the original track. The 'cloverleaf', as its name suggests, was a series of four 30 degree banked turns, each through 270 degrees in different directions so that none of the four overlapped each other, resulting in a four-leaf-clover pattern. All these patterns, when continuously repeated, covered the same area of ground, except that they drifted down-wind. On a typical test sortie, three or four of each pattern were executed. They were done in an area of good Decca coverage so that a reasonably accurate position of the Nimrod, during each pattern, could be determined and, like the en route navigation tests, the Decca position was compared directly with the INS derived position.

Inevitably, the search manoeuvres were more demanding on the stability of the INS gyros. Consequently, the positional accuracy of the on-board system drifted more noticeably than for the en route scenario. Nevertheless, after hundreds of hours of navigation system flight-testing and a vast amount of statistical analysis of the results, the accuracy levels achieved were accepted by the MoD.

During this period of navigation testing, I did some flying, in Nimrod XV226, on the MAD system. The first operational MAD had been installed on this aeroplane. It was housed in a long boom extending from the tail of the aeroplane. Besides having to assess the boom's effect on directional stability, some thorough flight-testing had been undertaken to ensure that the MAD was sited far enough aft so that the fuselage did not unduly interfere with its ability to detect submarines underwater. Inevitably there would be some magnetic interference from the Nimrod airframe, and a process had to be devised to compensate the MAD system for such errors. This involved a series of pilot-controlled oscillations of the flight path, separately in each axis: roll, pitch and yaw. It was important for the analysis that these oscillations were as close as possible to a sinusoidal profile of a regular period. The MAD engineers were not too happy with the accuracy being achieved by the pilots, and so a programme was introduced into the autopilot computer that would accurately 'fly' the Nimrod to follow the sinusoidal profiles. The on-board 'Digital' computer was programmed to use the information from these tests and automatically apply the compensation to the MAD's detected information. This autopilot system was eventually developed and installed in XV226 for assessment. If anything was designed specifically to induce vomit, this was it: a sideways-facing seat, no window to provide any visual external reference and being subjected to quite large oscillations, particularly in roll and pitch. I coped quite well. It was always easier for the pilots; they had a good view of the real world from the cockpit windows and could anticipate the motion. Test pilots were used to this sort of motion, but some were less keen; there were not many who willingly flew these MAD flights.

The production Nimrods were coming off the line and entering their production flight-test programme. Apart from the flight handling and performance checks of critical areas, the core on-board systems were a large component of this test schedule. As the Tac Nav was vital to the operation of the systems, each production flight test had to have a Tac Nav on board. I filled this task on many production Nimrods over a two-year period. It was always interesting to interface with all the other crew members, several of whom were specially trained in various search and detecting systems. Some of the live system testing was not feasible on routine test flights, so had to be simulated. The on-board 'Digital' computer was programmed to simulate the dropping of a pattern of sonar-buoys and these were displayed on the Tac Nav's CRT display. A programmed flight track to overfly these sonar-buoys was input to the system. The Tac Nav connected this to the pilot's flight director who would follow the track so that the sensor operators could pick up the simulated signals.

The objective of the production flight-testing was to demonstrate that each aeroplane's flight characteristics and systems functioning met the design specification. With the complex interface of systems on the Nimrod, some of our production flight-testing arguably went beyond this concept. One example was the MAD system compensation process. Checking that the MAD worked as a sensitive magnetic sensor and its information was interfaced with the other

HS Nimrod XV226 showing bomb doors open and MAD boom. (*Avro Heritage Museum*)

search systems was contractually as far as we needed to go, but having devised the error compensation process, it seemed sensible to complete this as well. For these tests, I moved from the Tac Nav seat to the system operator's console and soon made a name for myself as the only test observer who would readily volunteer for the flight on which the MAD compensation tests were being carried out. After the first batch of production aeroplanes had been delivered, the RAF was keen to accept responsibility for conducting some of the system testing themselves. One test on their list was the MAD compensation. It was not long after the first Nimrod delivery on which they were scheduled to conduct the additional tests that the RAF asked that we take the MAD compensation back into our pre-delivery schedule for the production aeroplanes. A little later, the RAF actually returned that Nimrod to Woodford for various reasons including the completion of the MAD test. Nothing was said, but we concluded that few, if any, of the RAF crew members had the stomach for it. I had a secure job.

Chapter 7

Restless for a New Challenge

At the end of 1971 and the beginning of 1972, things were beginning to change at Woodford. The production Nimrods were coming to an end and there were no new flight-testing projects in the foreseeable future. In addition, changes were being made to the organization of the technical departments. The Aerodynamics Department was based in buildings adjacent to the production assembly lines on the opposite side of the airfield to the flight test offices and hangars. Their management realized that there was a pool of people with potential design capability in the Flight Test Sections of the Flight Development Department. The solution was to incorporate our Flight Test Sections into the Aerodynamics Department. There was some logic in this, as the peaks and troughs of flight test and design workloads tended not to coincide, so this arrangement would make the best use of the available resources. The problem from my point of view, and that of other similar Flight Test Section personnel, was that the chain of approval signatories for our flight-test reports doubled in number.

At about this time, Doug Scard had replaced John McDaniel as the Flight Department Head. John McDaniel clearly had felt that he was ready for a new challenge. Under a complicated HSA contract with the Indian government, which was tied up with an agreement to allow 748s to be manufactured in India and the sale of the Hawk jet trainers, he was posted to India to manage the refurbishment of Russian MiG-21 fighters for the Indian Air Force.[1] Doug Scard realized that a large part of the technical analysis work of the Flight Department had become flight-systems orientated. This work was almost entirely done by the Flight Handling Section staff. The decision was taken to introduce a new section alongside the Handling Section and the Performance Section. It was called the Flight Systems Section. Obviously, Derek Bentley was the only contender to head up this section and I was pleased to become his deputy. Generally, this did not impinge on my flight test observer duties. After years of working and socializing together, we made a very good team.

There was a steady stream of 748s needing production flight-testing. In addition, Rolls-Royce had found they could squeeze a little more power out of the Dart engines and Avro (HSA as it then was) were keen to take advantage of this. In early

1972, we embarked on a short but very intense flight-test programme to ensure that the certification handling requirements were still met. Tony Blackman was again project test pilot. We completed this testing within one month, using about nineteen flying hours and culminating in the customary checks by John Carrodus.

At that time, the UK government had determined that all aspects of the safety of aircraft should be placed within one body. The Civil Aviation Authority (CAA) was formed for this purpose and the CAA's Safety Regulation Group took over the functions of the Air Registration Board (ARB).[2] John Carrodus and all the ARB staff, including those in the Flight Department, were now part of the CAA.

There was still a great deal of analysis and report-writing to do over the next few months. However, after that there would be no new flight-test programmes on the immediate horizon. Within Hawker Siddeley Aviation as a whole, there was a lot of ongoing design work with some divisions short of suitable staff. At Kingston in Surrey (originally part of the Hawker company before being subsumed into HSA), they were heavily into the Harrier (Jump Jet) design work as well as the Hawk jet trainer. The head of our Aerodynamics Department decided that I would be a suitable person to be seconded to Kingston to help them out. So for about two months I travelled to Kingston early on Monday mornings and back on Friday evenings. At first I made the journey in the Mini Cooper S, albeit mostly with the 997cc engine. I decided to replace my rusty old Mini Van with a second-hand Mk 1 Ford Cortina. This did the job and saved excessive mileage on the Cooper S.

I had no idea what to expect at Kingston. It was a very pleasant surprise to find myself involved in the mathematical modelling of the throttle control and engine response of the Hawk aeroplane. To minimize expense, Hawker Siddeley Aviation had decided to utilize an off-the-shelf fuel-control unit for the Hawk's Rolls-Royce Viper engine. The one chosen was from the Rolls-Royce Dart turbo-propeller engine, but no-one was quite sure if it would be sufficiently responsive for such a jet training aeroplane. Certainly, this was very new work for me, but I quickly realized how valuable my experience of setting up autopilot control laws would be. I had the use of a computer into which a variety of 'logic modules' could be plugged. These were chosen and arranged to represent the fuel-control unit system and the engine's response. The parameters for the control laws and engine responses were input into this mathematical model. There was some scope for varying the parameters within the fuel unit to best match the engine response characteristics and, on the face of it, a result was obtained that seemed to be potentially satisfactory.

However, this was a mathematical model and not a 'simulator'; there was no means for a pilot to make a direct assessment of the engine response to throttle movement. Having become quite adept at working with this computer modelling and with my experience of pilots' responses, I thought I would try to mathematically model a pilot! In flight, the throttle control is primarily used to control the airspeed. This may be to simply increase or decrease airspeed as well as increasing power to maintain a safe airspeed in a climb. So, taking the need to change speed as the trigger, I modelled what I thought would be a typical pilot response. The result looked quite good to me but the senior designer, quite rightly, thought that we should invite one of their test pilots to pass an opinion on this modelling and the engine response. I had not met any of the HSA fast jet test pilots, but the one who came along was a very pleasant youngish chap who was very interested in the modelling and looked carefully at the results. He thought that the interpretation of a pilot's reactions looked quite realistic and that the engine response would be acceptable if accurately represented. A good result; it was one of those days when my job satisfaction quotient was in the roof. The next step, to confirm that this fuel system would work satisfactorily on the Hawk, would be to incorporate the system into the Hawk simulator for assessment.

After that, I did not have much secondment time left, but I was given the task of making some aerodynamic calculations on the Harrier, in particular, to assess the adequacy of the aileron controls in normal flight.

Back at Woodford, there was more aerodynamics work on the agenda. At Hatfield (the ex-de Havilland part of HSA), they were working on the design of the HS146. It was a medium-sized four-engine commuter aeroplane, later known as the 146RJ (Regional Jet), and many of the later models are still operating in the 2010s. Some of this design work was allocated to Woodford, including the flying controls. Being a jet aeroplane, it would have been usual to design it with powered flying controls, but in the interests of keeping things simple and reducing costs, the ailerons were mechanically controlled. Although it was not a very high-speed aeroplane, it was challenging to design the ailerons and their supplementary control tabs to ensure that the pilot's control forces were reasonable in all normal operations. In addition, they had to be heavy enough at high speed to discourage large control movements that could be structurally dangerous. With powered controls, the pilot's control forces are determined by the 'artificial feel system' which can be programmed to use the airspeed, and other information, to increase forces with increasing speed.

On manual flying controls there can be up to three hinged tabs on the trailing edge of the control surface. They are there to manage the pilot's forces, for example, on the aileron. One is the trim tab which, as described earlier, allows the pilot, by use of a trim control, to balance any residual force on the control wheel,

HS 146 for which the author helped to design flying controls. (*Avro Heritage Museum*)

so that he can take his hand off the control. The second is a geared tab. As the name suggests, its movement is directly geared to the deflection of (for example) the aileron, but it moves in the opposite direction to aerodynamically assist the aileron in its direction of travel. This reduces the amount of force the pilot needs to apply by an amount that is proportionally effective at all speeds. The third type of tab is the spring tab. This works like a geared tab, except there is a spring in the lever that operates the tab. With increasing airspeed, the aerodynamic load on the tab increases, and the spring is compressed so that the amount of tab deflection is reduced. Hence, its contribution to reducing the control force is diminished, and so the pilot's control force increases with higher speeds: the desired outcome. To achieve the optimum control forces over the full aeroplane operating conditions, a good deal of work is required to determine the sizes of the tabs, the gearing ratio of the geared tab and the compression spring rate in the spring tab. This was my task for the 146 ailerons; back to serious analysis of stability and control equations. I was close to completing this when the project was cancelled due to the downturn in economy following the 1973 oil crisis. A few years later, in 1976, after the UK aviation industry had been through another 'rationalization' and when the economic situation was more favourable, the 146 was relaunched as the BAe 146, but I was no longer involved.

In the spring of 1972, one of my close friends from the motor rallying fraternity, Tony Bailey who worked for ICI, had acquired a sailing dinghy in some dubious part-exchange deal. He asked me if I knew anything about sailing and would I be interested in having a go with him?

My answer was: 'Very little, and yes!'

This turned out to be the start of much 'trial and error' and a totally new interest. We both read some literature on basic sailing; Tony had fitted a tow-bar to his Hillman Imp and off we went to the local reservoir, just to the south of Macclesfield. Fortunate that it was a warm sunny day as we spent more time in the water than actually sailing! However, although we were beginning to get the idea, after a couple of hours we were ready to call it a day. We made the same trip three or four more times and were beginning to avoid capsizing. Before the next attempt, Tony mentioned that in the equipment he had acquired with the dinghy, there was a spinnaker. We again consulted the instructive literature about sailing with a spinnaker and set off to try it out. It was a repeat of our first expedition; we were in the water more often than in the boat. Nevertheless, after a few more visits to the reservoir, we were managing the spinnaker quite well. Sometime later, I was recounting this story to a more experienced sailor and he asked what type of dinghy it was.

'A Merlin Rocket,' I replied, in total innocence.

He then proceeded to explain that it was one of the more 'high-performance' dinghies of its time and he expressed his admiration for our achievements, given our lack of experience.

Back at Woodford, I began to think about looking for a new job. Bill Horsley, who had left to join the ARB (before it became the CAA) as a performance flight test engineer a few years earlier in 1967, felt that I too should try to join the CAA. Every time there was a vacancy for someone on the flight-handling side, he gave me advance notice and encouraged me to apply. The first opportunity was in 1972. I sent in my application, but did not get as far as an interview. Bill told me afterwards there were two well-qualified contenders already known to the CAA and if neither had impressed at the interview, I would have been next on the list. A year or so later, they had another vacancy. This time the emphasis was for someone with particular experience of light aeroplanes. I did not believe I fitted the specification, but Bill again persuaded me to apply. This time I received a reply thanking me for the application and explaining that I had not been called for interview as they had two candidates with a very strong light aeroplane background.

At one stage I must have been feeling really desperate as I applied to one of the (then) still thriving Macclesfield silk mills for the post of chief engineer. The manager invited me for an interview and we chatted about my background and the responsibilities of the job.

He then took me around the works and, back in his office, he said: 'I know you could do this job, but I am not going to offer it to you because I know you will become bored very quickly.'

What a sensible person; it certainly did bring me back to reality.

My colleague and close friend Peter Morgan had recently left Woodford to join EASAMS (Elliot Aerospace and Advanced Military Systems), a company involved with the development of integrated navigation and weapon systems for the MoD. As with Bill Horsley, Peter also encouraged me to apply to EASAMS. I had an interview, but I did not feel that the salary they were offering was commensurate with a move to Surrey. They were not inclined to up their offer, so nothing came of it. Again, a good turn of fate because all the good engineers soon left EASAMS to set up their own companies. Subsequently, EASAMS became part of Marconi Electronic Systems and lost its purpose and separate identity.

In the interests of ongoing training and for a bit of light relief, the Woodford Flight Department had organized some dinghy drills at the local swimming baths in Bramhall. Compared with the experience in Plymouth bay, this was quite fun. However, we were using a multi-person dinghy that was more representative of the types of dinghy used on the 748 and Nimrod aeroplanes; it was a useful experience.

Next it was back to serious flight-test work. I was pleasantly surprised to be asked to be one of the project flight test observers on the Victor K2 tanker programme. Flying in one of the famous 'V' bombers was something special: a real treat. Although I had worked on several aspects of the Vulcan, there were more experienced flight test observers than me at the time and I never flew in one. The number one Victor observer was Alan Vincent. He was from Handley Page and had been enthusiastically taken on by HSA at Woodford when they were awarded the Victor K2 contract. Alan was very helpful and supportive to me. He fostered my understanding of the procedures and explained the difficulties likely to be encountered by a rear-crew member of the Victor. Unlike the pilots, whose parachutes were an integral part of their ejector seats, we in the back had to fit ours in the crew room before flight. With this on, walking across the tarmac to the aeroplane was very uncomfortable. Climbing aboard, up the steps through the entry hatch and into one's seat was quite a strenuous activity, especially when carrying all the paperwork and information required for the observer's task.

The Victor tanker project became an extensive and demanding period of flight-testing for me; I was feeling temporarily less restive. Initially, there was a significant programme of flight handling and performance checking. The main aerodynamic change from the B2 bomber configuration was the reduction in wing span by about 4ft. This same modification had been made to the earlier Victor K1 tankers when they were converted from the B1, so its effect was fairly well understood. Nevertheless, some of the Victor's basic flying qualities needed to be re-established before we could determine the effect of trailing the refuelling lines on the flying characteristics. Additionally, the MoD wanted the Victors to be able to refuel each other; this necessitated fitting a refuelling probe on the upper fuselage nose ahead of the cockpit window. This phase of the flight test programme progressed well, and we found no significant problems. However, as normal, the pilots tended to make use of the autopilot in between tests, and it was becoming apparent to them that it was not performing as well as expected. (The full saga of the ensuing autopilot flight test programme was documented in Chapter 1.)

Soon after we had completed the first phase of the Victor tanker development, in May 1973 another interesting programme came along. This time it was entirely under the responsibility of the Flight Systems Section and Derek Bentley was in charge. The RAF needed to operate the Andover military freighter into difficult remote airfields where the normal 3 degrees approach angle would be too close to the local high ground or other obstacles. To this end, the MoD contracted HSA to explore the feasibility of using steeper approach angles. The other feature of

HP Victor K2 tanker with all three in-flight refuelling probes deployed. (*Avro Heritage Museum*)

such airfields was that they were unlikely to have any ILS facility, so no matter what the desired approach angle, visual guidance would be required. An existing system known as VASI (Visual Approach Slope Indicator) was used for the trials. These VASI units were installed in a box that was readily transportable. They were sited at the side of the runway adjacent to the threshold and could be set to any desired approach slope within reason. The VASI unit comprised a series of lights that indicated to the pilot on approach whether he was above or below the set angle. Red lights indicated that the aeroplane was below the glideslope angle, green above and white at the correct angle. There were three lights of each colour, for redundancy, arranged in horizontal rows: the green row above the white row with the red row at the bottom.

For the trials, we had Andover XS606 and used the MoD airfield at Bedford, as we had done for the Viper-Shackleton testing back in 1965. As Derek's right-hand man, running the trials and doing the analysis was a large part of my task; I had to split my time between this work and the flight test observer duties. We had a second observer, Dave Gibbons, who did a lot of the flying. The programme required a large enough number of approaches at different angles to assess the ease with which it was possible to control the speed on the approach and accomplish the flare to a successful landing. As the approach angle increased, so the difficulty of the transition to the landing increased. It was anticipated that the consequence would be heavier landings or a longer flare. If the probability of heavy landings increased, it would introduce a risk of structural damage. If the distance from the threshold, where the flare was usually commenced, to the touch-down point increased then the total landing distance would be greater. This could be a significant problem considering the type of remote runways that were envisaged. It would have to be accounted for in the scheduling of landing distances. To achieve a good statistical set of results it was essential to have a number of different pilots to do the flying. We had a good mixture: our own Bill Else, three RAF pilots (Flight Lieutenants Millar, Ledwidge and Semark) and the CAA Test Pilot John Carrodus was invited to participate to add to the sample size and give his valuable opinion.

These trials were planned to take place over a period of about five weeks in three or four sessions, each lasting three or four days. It was, therefore, necessary to stay locally for several nights and we found an excellent pub not too far from the airfield. Not only were the food and beer good, but it had a dartboard as well as a skittle alley in a basement room. None of us was much good at darts but we felt that our skittle skills were improving; so much so that the landlord suggested we should make up the numbers in his team for a competition night at another local pub. I think we acquitted ourselves fairly well, but we certainly consumed a good quantity of ale; I have no idea who drove back to our pub that night.

Production Andover C Mk 1 XS606, used for steep approach trials. (*Avro Heritage Museum*)

To make a comprehensive assessment, we planned to use four different approach angles, 3 degrees being the norm and used as the datum for the other results, 4.5 degrees, 6 degrees and 7.5 degrees. All the pilots coped quite well with 4.5 degrees and managed the flare to touchdown with reasonable consistency, but 6 degrees was more of a challenge. Mostly they did reasonably well but the frequency of prolonged flares, slightly heavy landings and higher than normal touch-down speeds was of concern. When we increased the angle to 7.5 degrees we expected these issues to be more prevalent, but we came across another more limiting problem. In order to achieve the 7.5 degree descent angle, the pilots found they were spending most of the approach with the throttles fully closed. If the approach speed inadvertently became too high, it was almost impossible to slow the aeroplane down within a reasonable time. The inevitable conclusion for this aeroplane, the Andover, was that 7.5 degrees was an impractical goal and we took it off our test list. In an attempt to improve the 6 degree landing consistency, we tried another solution. We set up two sets of VASIs: one at 6 degrees but positioned well before the threshold, with the second set at 3 degrees positioned at the threshold. The idea was that the pilot would fly down the 6 degree path until he saw the second set of VASIs coming into the white zone when he would transition to the 3 degrees angle, following the second set. This was a more demanding approach exercise, but the touchdown consistency improved.

One other aspect, which I observed during these trials, was the variability of pilot response characteristics. By and large, they all achieved consistent results, but the way they flew was surprisingly variable. Some were gentle on the controls using anticipation as the key to their success. Bill Else was probably the most notable exponent of this technique. At the other extreme was Ron Ledwidge. At the first sign of a deviation in speed or flight path, he would make a rapid and quite large control input. As soon as he detected this was having the desired effect, he would immediately remove or sometimes momentarily reverse this control input. There was no doubt that this technique worked well for him. I referred to him as the 'high-gain' pilot; a nickname he took in good humour. It reminded me of my attempt to mathematically model a pilot's response during my secondment to Kingston and I thought how narrow my solution had been.

Back at Woodford, there were another few months with little flight-testing to do before we were ready to commence the serious business of assessing how easy it was to fly the Victor's flight refuelling probe into the drogue of another Victor tanker.

There was the analysis of the Andover steep approach trials to be completed. In addition, during this period I had been selected along with a chap called Reg Boor from the Aerodynamic Design Office to attend the Advisory Group for Aerospace Research and Development (AGARD) Flight Mechanics Panel conference on 'turbulence', held at Woburn Abbey. This was a two-day affair full of interesting presentations by very knowledgeable engineers and pilots from many parts of the world. I learned a lot about the different types of turbulence and their causes as well as some innovative ways of mathematically modelling turbulence.

At about that time I had another phone call from Bill Horsley giving me advance information that the CAA was again looking for another handling flight test engineer. This time, he said, the specification would exactly fit my experience and I should look out for the advertisement in a forthcoming edition of *Flight Magazine*. Sure enough, the advertisement appeared after a couple of weeks or so. Just as Bill had suggested, the job description and personal experience required were almost a perfect match for me and I duly sent off my application. I did not have a pre-prepared CV (curriculum vitae), but I cobbled something together and filled in all the sections of the application form to the best of my ability. Bill had told me that it could be a few weeks before the applications were reviewed by the CAA.

Before I heard anything more, I was back flying in the Victor tanker for what turned out to be an extended period over the next six months. There was a lot to do to satisfy the RAF and the MoD that the Victor would operate satisfactorily both as a tanker and receiver. Having one hose that extended from the rear fuselage and another under each wing, it had the flexibility to refuel one large aeroplane or two smaller, fighter-type ones. The initial Victor-tanker to Victor-receiver trials we could handle ourselves. These went well, with several RAF pilots joining our pilots Charles Masefield, John Cruse and Harry Fisher. One of the Victors used for these assessments was XL189. It was later to become famous as the Victor tanker that delivered the last of five refuellings to the Avro Vulcan that dropped the bombs on Stanley airfield during the Falklands War, XL189 itself having been refuelled three times on the way there and once on the return.[3]

It was quite an art flying the probe of the receiving aeroplane into the tanker's drogue, particularly in turbulence, but the pilots found that it was not just a matter of getting the probe into the drogue; it needed to be inserted with some force otherwise the fuel connection in the drogue would not open. Once we were happy, the RAF crews were keen to make their own assessments with a receiving Victor. With our pilots flying the tanker, the RAF provided a considerable variety of aeroplanes to assess their ability to connect with the Victor's drogue. In particular, among others, I remember the large transport aeroplanes such as the Lockheed Hercules and the BAC VC10, and in pairs, the English Electric Lightning fighters and the Royal Navy Blackburn Buccaneer strike aeroplanes. It was the difficulty of coordinating the availability of these RAF/RN aeroplanes with our tanker that added to the overall length of the trials.

During these three months, things were progressing with my application to the CAA. I had a call from Keith Perrin, the CAA Chief Flight Test Engineer, who was making a routine visit to Woodford to check some critical flight items on a 748 production flight test. He asked if I would be available after work to have a chat in a convenient pub. I was flying in a Victor that morning so was not involved in the 748 flight, but I readily agreed. Unfortunately my flight had been somewhat delayed and, in my haste to get ready to join Keith at the pub, I had accidently broken my glasses without which life was difficult. In particular, driving would be out of the question. Fortunately, I always kept a spare pair in my briefcase. However, some time before, I had them coated to use as sunglasses; a great idea and very useful they had been. They had thick dark rims and, with my hairstyle of the time and the dark glasses, people said I looked like Roy Orbison. It was OK driving, but when I went into the pub it was rather gloomy. I spotted Keith and went towards his table. I cannot imagine what his first thoughts were at seeing this strange sight; a potentially responsible CAA employee! I explained the situation as quickly as I could, and he seemed to take it in his stride. I don't remember any details of our conversation, but it seemed to go quite well and after an hour or so he departed for his home in Surrey. I remember thinking it was quite a long journey to be starting so late in the day. I was encouraged that he was prepared to do this.

Some weeks later, still in the middle of the Victor flying programme, I went to Redhill in Surrey for the interview. I was feeling quite confident and full of enthusiasm, but my interview technique had barely moved on from the university entry days. On the other side of the table were Dave Davies, the chief test pilot, Keith Perrin and a representative from the Personnel Department. When asked what attracted me to the job, I remember unduly emphasizing the pension and conditions rather than the interesting and challenging work. Keith had some well-thought-out 'what if' questions relating to flight-test scenarios to test my ability to come up with appropriate explanations. These were mostly situations of which I had experience and should have trotted out a good answer, but rather than the obvious, in some cases I tried to devise overly complicated explanations without much conviction. I could sense the disappointment of Dave and Keith. Finally, Dave made a closing remark about having noted I had applied three times and this would work in my favour.

With this comment, I relaxed somewhat and I said, without giving it any thought: 'I kept trying because I really like the idea of the job and believe I could do it.'

Their expressions turned into gentle smiles; this was the sort of enthusiasm they had been trying to induce from me for most of the interview. They thanked me for coming and said they would notify me within a week or so.

Quite a bit more than a week went by and I telephoned Bill. He was honest enough to tell me that they had one other person in the frame for whom they had a slight preference, but he was not able to give them his commitment. I later found out that it was Bob Wilson of Britten-Norman Aircraft based on the Isle of Wight. Britten-Norman were very reluctant to lose him and, additionally, he was having doubts about moving his family from the Isle of Wight. Britten-Norman offered him a post as engineering manager and that was too tempting for him to leave. Again, a bit of fate was working for me and I was offered the job.

Sadly, at that time, my father had become ill and the initial prognosis was not good. Now it was me who was having difficulty in making a commitment to move away from the Macclesfield area. The CAA had been my goal for some time; the salary and working conditions were more than I could ever have expected, but I had to give things more thought. There were few opportunities for promotion at Woodford because work was beginning to run down. However, I spotted an opportunity in the Flight Development Systems Engineering Department. The one project coming along was the upgrade of the Nimrod to the MR2 standard with huge advances in almost all the on-board navigation, submarine search and attack systems. I knew they were looking for new people and I approached the department head. He was quite keen to take me on, but he was not prepared to offer me the promotion I was looking for as he, probably rightly, thought it would seem unfair to some of his most experienced senior staff. I had run out of ideas.

My father pointedly did not wish his problem to influence my decision; also he was beginning to show signs of improvement which were confirmed by his doctor. I took the decision to accept the job offer. Subsequently he continued to improve, and I saw quite a lot of him over the next years.

★   ★   ★

Meanwhile, Tony Bailey and I were becoming ever more enthusiastic and experienced dinghy sailors. We had trailered the Merlin Rocket dinghy up to Anglesey in north Wales for a long weekend with the families. Sailing it off a beach into the waves was again the beginning of a new learning curve and time in the water! Fortunately, the weather was hot and sunny; it was just as well as we had no facilities for drying ourselves in our tents!

After this, we thought we should see if we could charter a large cruising yacht for a week's holiday. There were six of us in all: Tony, his wife, my first wife and me, with a couple of friends. Tony and I considered the possibilities. We realized that, with our limited experience, we would have to charter with a qualified skipper in charge. For some reason, Tony settled on an eight-berth catamaran with its owner as the skipper, chartered from Poole in Dorset. We had both enrolled on a ground-based yacht navigation course, hopeful of learning enough to be useful members of the crew. Because we had very limited budgets, we could not afford to eat out in restaurants, therefore we brought with us boxes full of assorted provisions for cooking simple meals on board. This clearly horrified the skipper, who apparently was used to his crew regularly taking him out for meals during the charter.

There was another factor that nearly caused the cancellation of the holiday. Tony had been involved in a serious car accident. He had a broken jaw and was missing some teeth. In order for his jawbone to mend, it had been wired together. However, to keep him fed, he could pour soups, fruit juices and drinks through his missing teeth! Despite this condition, Tony was determined to continue with the holiday; he was not going to give up on the plan. This was the second issue that horrified the skipper. Very reasonably, he was concerned that if Tony suffered from seasickness, he could choke to death! However, we pulled out our wire-cutters to demonstrate that, in such an eventuality, we were ready to release his jaw! Like me, Tony was an experienced rally navigator and, along with his recent sailing experiences, he was unlikely to succumb to seasickness.

The yacht was an Iroquois catamaran which, of its type, had an impressive performance. We sailed across the Channel, overtaking many other sailing boats on the same route, and visited Guernsey and Cherbourg. On our last night in Cherbourg, the skipper went out to join some of his friends in a French restaurant while we fed ourselves on board. The next morning, he was suffering from a serious upset stomach. He managed to struggle for a while to help us safely negotiate the exit from the harbour and retired to his bunk, leaving Tony and I to manage the yacht back to Poole. A great experience and a tribute to the skipper's confidence in the pair of us. He recovered sufficiently to guide us back into Poole harbour and onto his berth. The sailing bug had truly bitten us both and we each went on to separately charter sailing boats very many times over the following years. The experience we had gained from this first charter, together with the references provided by the skipper and some more extensive ground-school qualifications, enabled us to charter without needing a skipper from then on.

My last flight test from Woodford was in the Victor K2 tanker on 8 July 1974. I left Hawker Siddeley Aviation towards the end of August with some sadness, but I knew it was the beginning of an exciting new phase of my career.

Notes

1. McDaniel, *Tales of the Cheshire Planes* (1998), p.99.
2. Chaplin, *Safety Regulation: The First 100 Years* (2011).
3. White, *Vulcan 607: The Epic Story of the Most Remarkable British Air Attack Since WWII* (2006), p.18.

Chapter 8

From Industry to Regulation:
A Different Approach

Having accepted the invitation to join the CAA as a flight test engineer, I still had to solve the logistical problem of living close to the head office in Redhill. We had put the house in Bollington (Cheshire) on the market, but there was to be no quick sale. Fortunately for me, my ex-Avro friend Bill Horsley and his wife Sandra very kindly offered to put me up during the week at their house in Horsham. I was committed to commuting the 200 miles at the beginning and end of each week until the house could be sold and new accommodation found in the Redhill area. Little did I know how long this situation was to last.

So the day before I was due to start my new career, I drove down to Horsham in my Mini Cooper S to be sure of a timely arrival the next day. Bill drove me to Redhill the next morning and deposited me at the CAA reception in Brabazon House. The next hour or so will stay in my memory for ever.

Soon after I checked in at the reception, the personnel officer came down to greet me. First impressions can be very illuminating. She was a picture of elegance, very smartly dressed in a full-length skirt. She introduced herself as Ruby Kirton and immediately put me at my ease with her friendly demeanour. What an introduction to a new career. Totally beyond anything I had previously experienced and I hope my jaw had not dropped too visibly! She escorted me up to the Personnel Department and patiently and efficiently took me through all the administrative information and procedures I would need to be familiar with.

Many of these I had brushed up on from the information previously provided. I listened carefully to confirm my understanding of the main issues: the salary and career progression routes; the policy of first-class travel (it should be remembered that at that time, the option of business-class air travel had not been introduced and the cost differential between economy and first was very much less than it is today); the amount of leave; the pension costs and benefits. The background to the formulation of these arrangements was interesting.

A few years earlier, the UK government had decided that the Air Registration Board should be taken out of the Ministry of Aviation and be set up as part of a non-profit-making, self-governed entity, the Civil Aviation Authority, albeit with government-defined terms of reference and a government-appointed chairman. Importantly, it had to be self-financing. Although it was no longer a part of the Civil Service, many of their procedures were retained. In particular, the pension scheme was similar in its generosity, but with no direct government subsidy. It was set up, with a base fund, from the previous Civil Service system, proportionate to the number of staff transferred and, although the CAA made employer contributions, the individual staff member's contributions were quite large. I had reached an age where the comfort of having a good pension was an important consideration. Ruby had said that I would have the option of transferring my accumulated British Aerospace pension into the CAA system. This sounded like a good idea, so we agreed to look into this.

After being satisfied that we had covered everything in enough detail, Ruby then escorted me to the Flight Department, introduced me to some of the staff and showed me to my office. She pointed out, with genuine disappointment, that I would not be able to meet the department head and chief test pilot Dave Davies because he was on a test flight that day. However, the chief flight test engineer Keith Perrin was there. I had met Keith several times before, so this was an easy introduction and he gave me a very warm welcome. I would be sharing an office with Keith and the other 'handling' flight test engineers, Don Burns and Dave Cummings. Don I met the next day as he was on the flight test with Dave Davies. Dave Cummings was the light aircraft specialist although, as I soon discovered, we all got involved with just about all aeroplane types. There were two 'performance' flight test engineers, Andrew McClymont and Bill Horsley (both ex-Avro friends), who shared a different office on the opposite side of the corridor with the helicopter flight test engineer, Mike Smith, who I had met before on the 748 tropical trials in Asmara.

Keith introduced me to the other test pilots Gordon Corps, John Carrodus and Darrol Stinton; I would meet most of the other Flight Department staff at lunchtime. Within the Flight Department, there was another section known as the Flight Engineering Section, headed by Laurie Hall. Its responsibility was to keep up to date with any new technical advances in aircraft design, often with direct involvement with the industry, and formulate new requirements wherever necessary. Getting to know people from other departments was going to be important and Keith sensibly suggested that, at first, the best way to meet them was to be introduced as and when they came into our office. This worked very well because there was a regular flow of people wanting to have some discussion on certification issues of mutual interest.

Soon lunchtime arrived, and we all went off to a pub. Lunching at a pub every day was certainly a new experience for me. Back at Woodford, we only ever did this on special occasions such as Christmas or when someone was leaving. Here, it was an important part of the day. The CAA building, Brabazon House, did not have any canteen facility, so eating out was a necessity. To compensate for not having any lunching facility, the CAA issued all their staff with luncheon vouchers. These were not unusual at the time and all the local lunching establishments, including many of the pubs, would accept these vouchers in payment for the food (and quite often for the drinks as well!). There were five or six pubs within easy driving distance of Redhill that we tended to use on a random basis. Listening to the lunchtime discussions, it was soon obvious why they needed a new flight test engineer; the workload was such that there was almost always some staff away on flight test duties. Not everyone in the office always went along to these lunches; people sometimes had other commitments such as meetings or shopping. Nevertheless, there was always a good turn-out and it did not take me long to appreciate the enormous benefits. Exchanging views and experiences from recent flight-test activities was an invaluable way of learning from others and developing a consistency in the application of the CAA requirements and procedures. This was a 'working lunch' concept that I soon came to value highly.

There was no hurry for me to get into the air; I had a lot to assimilate. I was reasonably familiar with the flight regulations part of the British Civil Airworthiness Requirements (BCAR), but I found there were other important regulations on which I needed to be up to speed. There were operational requirements imposed on the airlines and charter companies, such as the accident data recorder (known as the 'black box' despite being coloured orange). Then there were the procedural and policy requirements. How to undertake the investigation of a new aircraft type for the ultimate issue of a UK Type Certificate was covered in BCAR Sections 'A' and 'B' (see Appendix II). Thereafter, each time a Certificate of Airworthiness was issued for a new or newly-imported aircraft, there were CAA checks to be performed and the Flight Department had their own specific procedures. I learned that there would be a compulsory CAA flight test for the first aircraft of a particular type on the UK Register, known as a 'series check'. This ensured that its flying characteristics in critical areas were consistent with those experienced during the Type Certification investigation. Subsequent new aircraft of the same type may also be subjected to the same flight test on a sampling basis; the frequency of such sampling was dependent on the perceived difficulty of achieving satisfactory results in the critical areas. Certificates of Airworthiness were generally renewed every two years, with mandatory inspections including a flight test. The schedule for this was simpler than the series check and was known as the Airworthiness Flight Test Schedule (AFTS). Most operators could carry out this C of A renewal flight test themselves using their nominated pilot(s) who were approved for this purpose by the CAA pilots. Occasionally, the CAA would carry out this test themselves for quality control purposes and sometimes for our pilots to maintain their 'currency' on the type.

During one 'working lunch' session a week or so after my arrival, Dave Davies enquired: 'John, how are you getting on so far?'

I replied: 'Quite well, I think, but I am very conscious that there is a great deal to learn.'

His response was immediate and encouraging: 'Don't worry; you will soon find it all falls into place.'

There was one significant difference between the industry flight testing and the certification testing carried out by the CAA: any aircraft in which the CAA would be conducting flight testing should, and almost invariably would, have been subjected to those same tests by the manufacturer during their certification programme. In theory then, the risk level would be lower. A comforting thought, but I had learned not to be complacent.

After a couple of weeks, I was ready for some flying. The first opportunity was a series flight test on a Rockwell Commander 685 at Staverton near Gloucester, with Gordon Corps. This was a small US twin-piston-engine business aeroplane and the 685 was a recently-certificated variant. It was my first exposure to this class of aeroplane and it was interesting to see how responsive they were compared to the larger types with which I was familiar. We went through the flight tests in our Flight Test Schedule with Gordon gently leading me. The only difficulty we had was the measured climb performance. In all the series and C of A Renewal Schedules, we included a five-minute measured climb check, in the en route configuration, to confirm that the scheduled level of performance was being maintained. With many smaller aeroplanes, particular those of US manufacture, the CAA had sometimes found that the scheduled rate of climb was a little optimistic. This Rockwell climb performance was rather marginal, but we always allowed some tolerance from the scheduled value to account for atmospheric variations, so we decided we could accept it. However, we made a note to closely monitor other future test results of this type. A notable feature of these Rockwell types was the slightly unusual nose-wheel steering. Larger aeroplanes often had a separate 'tiller' to control the nose wheel on take-off until the airspeed was fast enough for the rudder to take effect. The other method was to connect the rudder pedals directly to the nose-wheel steering as well as the rudder; the nose wheel was disconnected as soon as it was off the ground. On the Rockwell Commanders, the nose-wheel steering was achieved through the initial deflection of the wheel-brake pad on the

Rockwell 685, G-BBHA; the subject of the author's first test flight in CAA. (*ABPic, Peter Davis*)

rudder pedals. This signals a hydraulic actuator that deflects the nose wheel. It could be difficult to determine the point where the nose-wheel steering actuation ceased and the wheel braking commenced. This was a technique that took a bit of getting used to, particularly on take-off, but Gordon had previous experience and managed it well. After our flight, we met with the local CAA surveyor from the Bristol office and went through our results. This completed the information he required, and he could issue the Certificate of Airworthiness.

The CAA had an extensive network of area offices sited at strategic locations throughout the country, including Scotland and Northern Ireland. These were staffed by CAA surveyors with maintenance and/or manufacturing expertise. It was their responsibility to inspect individual aircraft and its documentation before issuing the C of A. As part of this process, they needed to know the results of the series of AFTS. Inevitably I would get to know many of these area surveyors over the following years and appreciate their involvement and cooperation.

I soon realized that, in contrast with the working practices I was used to at Woodford, in the aeroplanes we were flight-testing at the CAA, there was invariably no purpose-built console from which to work. Frequently, the only flight instruments available to read were those on the flight-deck. On larger transport aeroplanes, there were usually spare (supernumerary crew) seats in the flight compartment. In smaller types like the 748 there was the so-called 'jump seat' that folded out from a compartment behind the pilots; when extended it fitted just aft of and between the two pilots. With no 'desk' to write on, my well-used clipboard was invaluable. As always, I needed ready access to the usual stalling speed and operating limitations, but for each different aeroplane type, I had to extract the data from the relevant Aeroplane Flight Manual. For every aircraft type on the UK Register, the CAA Flight Manuals Section had a copy of the relevant CAA Approved version. Naturally, they were reluctant to let their 'master copies' out of their sight, so the Flight Department had their own duplicate copies for most of the more common types. It was generally easier to copy the relevant charts and limitations from the Flight Manual and to attach them to my clipboard than to take along the whole Flight Manual. If we needed to know anything else, such as specific emergency procedures, we could consult the Flight Manual which was always required to be on board the aeroplane. For each individual aeroplane, it was also important to know of any calibration errors in the pilot's flight instruments.

All the CAA test pilots carried with them a 'spring-balance' with which they could measure the elevator and aileron control forces they needed to apply in certain tests. In common with most experienced test pilots, they were all pretty good at estimating the control forces they were applying. However, when undertaking precise certification tests where the forces were critical, in the absence of any real-time instrumentation read-out, it was imperative to know the force being applied with some confidence. I think it was John Carrodus who developed the design for this spring–balance for flight-test work. It consisted of a tubular barrel, about 4cm in diameter, serrated to give a good grip, with a rod longitudinally through it acting on a spring within the barrel. At one end of the rod was a hook that could be looped around the control-wheel for measuring the elevator 'pull-force', and at the other end there was a V-shaped attachment used for pushing. In the side of the barrel was a slot from which a pointer, attached to the spring, protruded and its position was read off from an adjacent scale on the barrel. It could also be used for measuring aileron control forces, but it was often more difficult to apply the force in the correct direction and simultaneously read the scale. Gordon used this device very effectively to check the critical control forces on the Rockwell 685 flight test.

I also found that it was customary for flight test engineers to fly with the approved airline pilots on some of the C of A Renewal air tests. This had the advantages of relieving our test pilots from excessive routine work, while continuing the monitoring of the aeroplane type in service and keeping an eye on the competence of the airline pilot for these testing tasks. In my early months, I did quite a lot of this work, particularly on the HS (de Havilland) Tridents of British Airways (BA). BA had several competent pilots for this task on all their aeroplane types, as well as an excellent back-up team of flight test observers and analysts, some of whom I would come to know very well.

It was becoming obvious that the housing market was going through a difficult time and any chance of a quick house move south was looking unlikely. I was not happy adding so much mileage to my Mini Cooper S with the weekly commute to Cheshire and my old Ford Cortina was, well, 'old'. It was the time of the first serious 'oil crisis' and petrol prices were increasing rapidly. There was a sudden glut of high fuel-consumption second-hand cars on the market at ridiculously low prices. Conversely, cars with a reasonable consumption were in high demand and prices were increasing. It did not take much mathematics to work out that I could afford a lot of extra fuel from the price difference to justify purchasing a cheap fuel-guzzler. I bought a Ford Capri 3000 for a ludicrously low price; what a car to enjoy driving!

My first CAA flight test with John Carrodus was, coincidently, in a 748. The Civil Aviation Flying Unit (CAFU) had installed the new Ground Proximity Warning System (GPWS) into one of their aeroplanes for assessment and development prior to CAA approval. The incidence of pilots inadvertently flying into the ground in bad visibility had become one of the main causes of fatal accidents. This system provided a means to warn pilots of the aeroplane's dangerous proximity to the ground. It had a radio-altimeter that input the height above ground into a computer, which was also fed with the position of the aeroplane's wing-flaps and undercarriage. If the undercarriage (landing gear) was down, it knew the

The author's Ford Capri 3000 at Biggin Hill for flight with Darrol Stinton, 1974. (*Author*)

aeroplane was probably on approach to land; if only the flaps were down, it was in a descent for the airfield or on a climb after take-off. In all cases the GPWS was programmed with minimum heights for safety or dangerous rates of closure to the ground and would give voice warnings to the flight crew. These were 'Too low, gear' if the aeroplane was well below the glideslope on an approach to land; 'Too low, flaps' if the aeroplane was too close to ground in the vicinity of the airfield; and 'Pull up' when there was perceived to be an immediate danger in any configuration. The system worked quite well, the only problem being the voice synthesizing which, at that time, was in a very early stage of development. As I remember, the main difficulty was the pronunciation of the 'l' and 'u'. The 'Pull up' sounded more like 'Poo oop'! Nevertheless, the intention was obvious and the subsequent introduction into commercial airline service of this system has undoubtedly prevented many accidents. It did not take long for advances in voice synthesizing technology to improve the diction.

In the absence of John Carrodus on other work, I was to monitor some of these GPWS assessment flights on the 748 with the CAFU pilots. This testing included some deliberate descents into the Welsh hills. This all went extremely well, but afterwards I had to write my first CAA Flight Test Report. With the pressure of other work, I had to complete it quite quickly and, in the absence of John Carrodus, I asked Dave Davies, being department head as well as chief test pilot, if I should wait until John's return for him to approve it.

Dave's reply was immediate: 'You wrote it; if you believe you have included everything in it and it is all correct, then you sign it.'

What a breath of fresh air compared with the report approval process back at Woodford.

The amount of report-writing and the urgency to get them completed rapidly improved my 'English'. Out of necessity, and to my delight, I found that my vocabulary was increasing and I became able to formulate succinct sentences quite rapidly.

A couple of weeks later, I had my first flight with Dave Davies; a C of A Renewal air test on a BAC VC 10. I was more than a little apprehensive, but everything went smoothly. I was particularly impressed with how he was able to remember specific airspeeds for many of the tests, which made my job easier. This flight was closely followed by similar flights with Dave on a BAC 1-11 and a Boeing 707. He was totally in control and, bearing in mind that he had not flown any of these types for many months, he flew them with competence and precision.

My next flight was with Gordon Corps: this time it was a de Havilland Canada DHC-6 Twin Otter. It has a fixed (non-retractable) undercarriage, and this one was fitted with especially large tyres intended to provide some extra buoyancy in the event of an emergency landing on water. We carried out a normal airworthiness flight test, with a few extra checks to determine if the big tyres had any measurable effect, particularly on the directional stability and climb performance. Gordon recounted an interesting design conundrum during the early design of the Twin Otter. It has the wing mounted on top of the fuselage, and to simplify the mechanical linkage it was decided to mount the engine throttle and propeller pitch controls in the roof of the cockpit rather than the normal location on the central console between the pilots. However, there is an accepted convention for the operation of switches and levers in the flight compartment. It is important that there should be no confusion, particularly as switches and controls may be mounted in various locations on the central and side consoles, the forward instrument panel and the roof panels. The accepted convention is that when the pilot sweeps his hand forward from the rear edge of the central console, upwards over the main instrument panel and backwards along the roof panel, this is the direction that will move any switch or lever to the 'on' position. Normally, this convention works well. For the design of the Twin Otter throttle levers to comply with this, they were arranged to move backwards to increase power. This was so unnatural for pilots who were conditioned to move throttles forward to increase power that very early in the initial development programme a concession to the convention was convincingly demonstrated to be necessary. The direction of movement of the throttle and propeller pitch control levers was reversed. On a worldwide fleet of about 1,000 Twin Otters, it was a decision that has never been challenged.

During these early months, I also began flying with Darrol Stinton. His flight-testing speciality was light aircraft, defined as aeroplanes with a maximum take-off weight of 6,000lb (2,730kg). Most of these on the UK Register were imported, predominantly from Piper, Cessna and Beech in the USA or Socata and Robin in France. The CAA had reciprocal agreements, with the Federal Aviation Administration (FAA) in the USA and the *Direction générale de l'aviation civile* (DGAC) in France, to allow the import of each other's aeroplanes in this category without a comprehensive design investigation. However, there were always a few items from the different national requirements that had to be checked. For example, the CAA had a specific requirement for the duration of the emergency electrical power supply to the primary flight instruments; this often necessitated a larger battery. There were also some flight items such as stall-handling, spinning and the measurement of climb performance (for the reason mentioned earlier) which necessitated a series air test. We would take this opportunity to check other things such as the emergency means of lowering the undercarriage where

DHC-6 Twin Otter G-BTWT with Gordon Corps (under the wing) at Biggin Hill. (*Author*)

applicable. Occasionally, Darrol would also arrange to fly the C of A Renewal air test on a type he had not flown for some time or one that we had found to have a history of marginal climb performance or dubious stalling characteristics.

Normally, my flight test engineer colleague Dave Cummings would accompany Darrol on such tests, but Dave was spending a lot of his time on the certification evaluations of the larger aeroplanes, above 6,000lb but less than 12,500lb (5,700kg); mostly manufactured by Cessna and Beech in Wichita, Kansas. Darrol was very pleased to have another flight test engineer to share the flying and we quickly struck up a good relationship. Here again my well-developed constitution was a great advantage, particularly during the spinning assessments. Spinning was one of the recognized 'high-risk' tests and we always wore parachutes. The manufacturer had to specify the best actions for spin recovery which invariably required the application of full rudder in the opposite direction to the spin; in some cases, movement of the elevator or aileron control was also specified. Darrol often came to ask me to join him when spinning was on the schedule even though Dave Cummings was available, as he knew that Dave was not overly keen.

The variety of aeroplanes Darrol had to deal with was amazing. In my first few months we flew an old Luscombe 8F tail-wheel aeroplane, new variants of the Mooney M20, the Piper PA 34, the Cessna 150 (semi-aerobatic version), the

Luscombe 8F, G-AFYD at Thruxton. (*Author*)

Cessna 336 clearly showing the 'push-pull' engine configuration. (*ABPic, Ad Vercruijsse*)

Rockwell 112 and a Socata Rallye 150, as well as the twin-engine Cessna 336 and 337, a Piper PA 23 Aztec, a Cessna 310 and an Italian Partenavia P68. The Cessna 336 and 337 had one engine in the nose and one mounted behind the cabin with a pusher propeller. The tailplane was mounted on twin booms extending from the wing trailing edge. With the occupants sitting in the cabin, which was effectively a 'sound-box', between two piston engines, it was certainly one of the noisiest aeroplanes in which I have ever flown. After each such flight test, we always made contact with the local CAA surveyor to report our results.

This set the pattern for the first year or so, flying in roughly equal measure with each CAA pilot on just about every aeroplane type operated by the UK airlines.

During my first months, I found that the CAA Flight Department had an additional responsibility I had not expected. Flight simulators were becoming an important tool of the airlines for training their pilots; this was much cheaper than using aeroplanes.

The airlines were pushing the CAA Operations Division to allow a higher percentage of simulator time to be credited as part of the total required pilot training time. For this to be legitimate, it was imperative that the simulator flying characteristics were closely representative of the actual aeroplane. To this end, the Operations Division was aware that the CAA Airworthiness Division test pilots and engineers had direct experience and important quantitative data defining the flying qualities of all the UK-operated aeroplane types. Consequently, they delegated the Flight Department to make the necessary assessments of all the simulators used by UK airlines.

At that time, in the mid/late 1970s, computer-generated moving visual displays were just becoming credible enough to be widely used in the latest generation of flight simulators. These newer simulators also had quite convincing motion systems. The flight-deck cabin, based on an actual section from the real aeroplane, was mounted on top of the hydraulic jacks which moved the cabin in pitch, roll and yaw in response to the pilot's movements of the flying controls. Because of limitations in the travel of the jacks, the one aspect of the motion that could not be simulated realistically was acceleration (g forces). Short-period forces like the bump on landing could be represented, but continuous forces in manoeuvres were impossible. The flight-deck windows were replaced with TV screens depicting the outside world. The simulator operator would have a small choice of airports within their local environment that could be set into the simulation. The processing speed of computers seemed miraculous at that time but, by modern standards, they were extremely slow. There were a very limited number of geographic data points that could be moved rapidly on the screens in response to the computed motion of the aeroplane. For this reason, only night scenes were initially possible, representing the runway and airport lighting together with limited representations of road and urban lights.

Even though the major airlines could afford to have these state-of-the-art simulators for their newest aeroplane types, there remained in use, for older aeroplane types and smaller airlines, many earlier technology simulators. Before computer-generated displays, the best technology was to have a TV camera moving over a three-dimensional model of an airport and its environment. These models contained a lot of detail and, by their very nature, took up a considerable area in the simulator facility. The camera moved over the artificial terrain and rotated in pitch, roll and yaw in response to the computed movement of the aeroplane. The camera's image was displayed on a TV screen in the pilot's windscreen; usually in colour and quite convincing except for the mechanical system controlling the camera movement often being a little jerky. There were versions of this system in use both with and without motion systems. Each had their uses as training aids, but those with motion clearly had the potential to be granted the most credit as part of the pilot's training flight time.

At the bottom of the simulator capability, and still in use by some airlines for older aeroplane types, were those with no motion and no visual display. The Vickers Viscount was a good example. Effectively, the pilot could only use these simulators to fly in full Instrument Meteorological Conditions (IMC). However, it was still important that the response of the simulated flight path to the pilot's control movements and forces was representative of the actual aeroplane for even minimal training credit to be considered.

Overall, the assessment of simulators was a significant part of the Flight Test Department activity; we had to keep a record of the relevant numerical data regarding the handling and performance characteristics of all larger commercial aeroplanes. For each of the test pilots and flight test engineers, we probably averaged one or two simulator assessments every month. These simulators, particularly the modern ones, were expensive and generally only the larger airlines with a big fleet could justify the investment. Hence, there was a good trade in hiring out time on simulators to other airlines. The Flight Operations Department also had a list of simulators not owned by UK airlines, contracted for use by some UK operators. This meant some overseas visits for the test pilots and engineers; for me, I remember simulator visits to Aer Lingus in Dublin, Scandinavian Airlines in Oslo and KLM in Amsterdam.

During 1975, the CAA was planning a visit to Russia to evaluate the Yak 40 for potential UK Type Certification. The Yak 40 was a small commuter passenger aeroplane with three engines in the tail, similar to the configuration on the Trident. This was the first commercial aircraft type from the Soviet Union to be offered for sale in the West and the CAA obviously had no experience of dealing with Soviet countries. Advice was sought from the UK government Foreign Office. They emphasized two particular issues to be addressed. Firstly, take a translator in whom you have confidence who understands aviation terminology. The Russians will want to use their own translators, but you will never know whether you are getting an honest or a 'political' answer. Secondly, always be aware that they will try many ways to put individuals from the CAA team in compromising situations. All this seemed a bit 'cloak and daggerish' but it did turn out to be realistic advice.

The CAA chose a suitable translator from a UK university and his visa application was submitted along with the list of CAA team members. Not long afterwards, the message came back that the translator's visa was refused; a response anticipated by the Foreign Office. The next step in the 'game' was for the CAA to refuse to make the visit unless the chosen translator could go too. After a couple of further rejections of unacceptable alternative solutions suggested by the Russians, the translator was finally granted a visa and the visit took place. The Flight Department team members were Dave Davies, Don Burns and Bill Horsley. Once there, the

Russian Yak 40, the subject of a UK CAA evaluation with several interesting stories. (*ABPic, Erik Frikke*)

team members commenced their investigations of the Yak 40 in their individual areas of expertise. As was customary, before dinner they all met together in one of their hotel rooms to discuss areas of difficulty anyone had encountered during the day. The next morning back at the Yak offices, when the CAA raised these issues, to everyone's amazement the Yak engineers had produced pre-prepared responses to each one. This was clearly only possible if they had overheard the CAA team's conversations by bugging the rooms. There was a total lack of any embarrassment on their part. Also, as predicted by the Foreign Office, the CAA team members all experienced a knock on the door from a charming young lady offering her services, which were equally charmingly declined.

Although the CAA team found the Yak 40 to have some significant problem areas, the main insurmountable difficulty was the lack of any independent civil airworthiness authority or any published requirements for civil aircraft in the Soviet Union. They had some military aircraft requirements which they also used for civil aircraft and each manufacturer had a 'Design Bureau' who would confirm the compliance with these requirements. Once the Design Bureau expressed themselves satisfied that the requirements were met, the certificate was issued by a high-ranking military officer. The biggest difficulty with this for the CAA was that the whole system of certification lacked any independent oversight as the personnel in the Design Bureau were the same engineers responsible for designing the aircraft!

One amusing situation that arose and served to illustrate the different Soviet approach was the subject of 'ditching'. In the West, it was a requirement to demonstrate that the aeroplane could be safely landed on water in the event of a total loss of engine power (a ditching). This was normally accomplished using a specially-constructed model that was 'flown' into a tank of water. This was not a requirement in the Soviet Union and nowhere did they have any facility for accomplishing such a test. The Yak proposal was to demonstrate a safe ditching by landing an actual Yak 40 on water; seemingly they had several unsold airframes and they thought this would be their cheapest option. They even invited Dave Davies to fly the test; he declined! The CAA certification never progressed so there was no necessity to further this proposal.

Late in 1975, my first wife and I, with the two children (5 and 7 years old) were finally able to move house to Sussex. I remember being amazed at how little time it took them to lose their northern (Cheshire) accents!

It was not long before the amount of my travelling all over the country was sufficient for me to qualify for a company car. There was not much choice available. For my grade, it was a Talbot (ex-Hillman) Avenger. There was a choice of colour, and I discovered that the estate version was an option if you paid the small difference in cost. I chose a red estate car; not surprisingly, my colleagues referred to it as the 'fire truck'. It gave me the opportunity to get rid of my Ford Capri 3000, which I had enjoyed driving very much but it was developing several reliability problems; not helpful if you regularly needed to be somewhere by a specific time.

In the winter of 1976, I had my first exposure to a Type Certification project; 'exposure' being an appropriate description for the cold weather 'route-proving' trials of the Short SD3-30 commuter aeroplane. In a similar way to tropical trials, it was essential to ensure that any new aircraft type could operate satisfactorily at the minimum operating ambient temperature specified in the Flight Manual. Just as for tropical trials, it was customary to have a member of the CAA Flight Department on the team to monitor the results. The SD3-30 was a small transport aeroplane seating up to thirty passengers. It had a square-section fuselage, a high wing, twin fins and was unpressurized. It was designed to have low maintenance costs.

In early January 1976, I had a 'briefing' visit to Short's at Belfast with Andrew McClymont. Andrew was the CAA's nominated performance flight test engineer

Short SD3-30, G-BDMA, in which the author participated in 'cold weather trials'. (*ABPic, Michael West*)

for Short's aeroplane types and would have been the first choice for this task, but he had other high-priority work. He came along to introduce me to Short's chief engineer as well as their pilots and flight test engineers. The planned itinerary was discussed and agreed. It involved an intense four days of flights from Belfast to Scandinavia, up into the Arctic Circle, back to Belfast and then around the Scottish Islands.

At the end of January, I was back in Belfast ready for the trials. After lunch on 31 January, we departed for a short refuelling stop at Aberdeen and then on to Oslo for the first night stop. We left the aeroplane outside overnight to give it a testing 'cold soak'. The next morning, everything worked OK and we left northwards for a stop at Trondheim, up into the Arctic Circle to Bodø, back to Trondheim and Oslo. This was a total of six hours and thirty minutes with no issues. The next day brought the only significant problem of the trip. After take-off from Oslo, the undercarriage would not retract; we completed a circuit and landed straight back. The Short's ground engineers, who we carried with us as part of the crew, soon diagnosed a frozen linkage which was easily thawed out and off we went again. Just the kind of thing that needed to be added to the ground checks on normal operational service in cold conditions.

On the subsequent taxi out for take-off, we heard over the radio in a very cultured Scandinavian accent: 'Beauty is in the eye of the beholder.'

Although the voice did not identify itself, it obviously came from the captain of the Scandinavian Airline DC 9 following us on the taxi-way; clearly a reference to the functional square shape of the SD3–30!

We then flew to Stavanger, Aberdeen and back to Belfast. The final day was a very interesting circuit of short hops to the Scottish Western Islands, some of which had grass-surfaced runways: Islay to Benbecula, Benbecula to Stornoway, Stornoway to Benbecula, Benbecula to Machrihanish, and Machrihanish to Belfast. With only one difficulty encountered throughout the trial, the subsequent debrief was quite short and I was off back on the late evening flight to Gatwick.

Back at Redhill, another report to write and some more routine test flights: a Boeing 707 C of A Renewal air test with Dave Davies and a spate of Piper PA-23 Aztec flights with Darrol. A demanding and satisfying first eighteen months of a new career and a feeling that I was beginning to earn my keep.

Chapter 9

Tiger Moth to Concorde

The CAA's Light Aeroplane Test Pilot Darrol Stinton's flying responsibilities were frequently very demanding. Sometimes he had to make flight tests in old aeroplanes and sometimes in home-built aeroplanes; often, in both cases, they were single-seat types and, therefore, he had to manage without a flight test engineer or another pilot to assist him. Reliability was not a strong feature of these types and, inevitably, engines failed from time to time. Anyone who has been a private pilot of single-engine aeroplanes will know that a key part of the training covers dealing with an engine failure. The first action is to stabilize the aeroplane at a safe speed for minimum rate of descent and then look around for a suitable field for a forced landing. Then, if there is time, check possible reasons for the failure, such as fuel selected from an empty tank. If the reason can be detected and corrected, a restart can be attempted. If this is not possible, success is entirely dependent on a good choice of landing site and a carefully-judged flight profile to arrive at the selected touch-down point, heading into wind at the right speed. Without an engine, this is a very difficult task.

During his career at the CAA, Darrol found himself in this situation numerous times. It is to his enormous credit that, in every case, he identified a suitable landing site and made a successful landing without any significant damage to himself or the aeroplanes. My only flight with Darrol where this possibility had been a close call was in an old Second World War Lysander. We had taxied out ready for take-off but found that, during the engine power checks, it was not possible to achieve full rpm. We taxied back. It turned out to be a carburettor problem and we gave up and returned home. On a later occasion Darrol, flying on his own, had an engine failure in a Tiger Moth. Again, he had successfully landed it without damage, and he asked me if I would like to join him when he returned the aeroplane to its home base after the engineers had fixed the problem and prepared it for the flight. My worry was that my extra weight would compromise the ability to get airborne, but Darrol was happy that the field he had chosen for the landing, near Tonbridge in Kent, was smooth and very large. There was no technical justification for taking part in this; it was really a 'jolly'! We were driven to the field, climbed on board, started the engine and took off successfully. The short flight back to the Tiger

De Havilland Tiger Moth. (*ABPic, Rinze de Vries*)

Moth's home at Redhill aerodrome was my first in an open cockpit and I enjoyed every minute. Snoopy and the Red Baron came to mind!

A few days later, I was again flying with Darrol. This time it was a Helio Courier H-295 from an airstrip on a farmer's field near Farnham. The farmer was an enthusiastic private pilot. This was his own aeroplane which he used to 'commute' between his farms. Because there were only short grass airstrips to use for take-offs and landings, he needed a very low-speed aeroplane that could be operated from such strips. The Helio Courier was ideal; it had a powerful engine and a high wing with substantial high-lift flaps. He always asked Darrol to do the bi-annual C of A Renewal flight test whenever Darrol was available because he was not experienced enough to do it himself and he had little confidence in the other pilots who we had approved to carry out such testing.

There was some crosswind for the take-off and I sensed that the aircraft was quite a handful to control, but Darrol managed it well and after a very short run, it leaped into the air. The tests went well but we were operating close to the Gatwick

Helio Courier H295, G-BAGT; the aeroplane in which we inadvertently entered the Gatwick control zone backwards! (*ABPic, Ron Roberts*)

Airport Control Zone, so we had to continually check our position as the wind was quite strong from the south-west and blowing us towards the zone. The last item on our flight programme was the stall testing. Darrol pointed the aeroplane away from the zone heading in a direction back towards the farm field. We carried out a series of stalls in all configurations in fairly quick succession. We had just about finished when we received a call on the radio from the Gatwick controller, telling us that we had entered the zone. Somewhat puzzled, after a bit of thought it dawned on us that the wind speed was higher than our average speed during the stalling and we had entered the zone backwards!

Gatwick subsequently filed a violation for the unapproved entry into their zone which found its way onto the CAA Chief Test Pilot Dave Davies' desk. Dave called Darrol into his office for an explanation and, with some embarrassment, Darrol told him what had happened. Dave was quite amused and, since Darrol was clearly trying to fly away from the zone, Dave used his influence so that no further action was taken. It was most probably the only occasion of an aeroplane entering the control zone backwards!

One of the 'perks' of a flight test engineer was a subsidized programme of training for a private pilot's licence. This was justified because it would be a clear benefit for us to have the first-hand knowledge and skills of flying aeroplanes. I took up this opportunity, as soon as I was able, with great enthusiasm. When I received approval to start the training, it was the beginning of winter. The nearest training airfield was at Shoreham but, at that time, it still had a grass runway. My colleagues in the Flight Department warned me that this runway was regularly waterlogged after heavy rain. It would be more productive to go to an airfield where there was a hard runway; Goodwood near Chichester was the obvious choice. Trying to fit in time for flight training with work was difficult enough without having to worry about the state of the runway on your free day.

The aeroplane used for training at the Goodwood Aero Club was the French-built Robin 200/100; a pretty-looking low-wing two-seater and a pleasant change from the ubiquitous Piper Cherokees and Cessna 150s normally used for training.

One of my initial difficulties when learning to fly was the rudder pedals. I knew very well that pushing on the left pedal moved the rudder to the left, causing the nose to move to the left. However, for some reason, I did not find it instinctive. As soon as I started taxiing, it was rapidly obvious which way to push the pedals to control the nose-wheel steering and after a few flights, I was beginning to use the correct pedal more instinctively. In a quiet moment, I reflected on the reason for this difficulty and remembered my childhood days when we made 'trollies' out of wood and old pram wheels. The front wheels were fitted at either end of a piece of wood (the front axle) which was attached to the trolley structure by a single bolt in the centre, acting as a pivot. The occupant sat on the trolley with one foot at each

Robin 200-100 in which the author learned to fly, 1975. (*Author*)

end of the 'axle' to control the direction of travel. To turn left required the right foot to be pushed forward; the opposite of an aeroplane! I mentioned this dilemma of mine to Darrol Stinton one day.

He gave it some thought for several seconds and replied, partially in jest, I thought: 'I wish you hadn't told me that; I am going to try and forget it!'

There was an annual limit on the number of flying hours for which the CAA subsidy would be given but, on top of this, there were additional hours allowed for the Private Pilot Licence (PPL) training which meant that the licence could be acquired within a year. However, flight-testing commitments took priority and, in my case, it took nearly two years. I managed my first solo flight after ten hours of tuition. This was about average, but it took me a total of forty-two hours to gain my private pilot's licence.

Several months after I joined the CAA, it was decided that the ever-increasing workload necessitated an additional test pilot in the department. We had been using Dave Glaser, a retired test pilot from British Aerospace at Weybridge, on a part-time basis. His experience on the Vickers Viscount, the BAC 1-11 and the VC 10 was very valuable, and I flew with him a few times. However, a permanent solution was needed and a few months later Nick Warner was recruited. Nick had been in the RAF flying Hawker Hunters and the Harrier jump-jet. He had completed the test pilot training at the Empire Test Pilot School (ETPS) at Boscombe Down. After leaving the RAF, he joined British Aerospace at Warton, flight-testing a variety of 'fast jets', before applying to join the CAA. Like me, he had to spend his initial time assimilating the CAA requirements and procedures before actively participating in the flight-testing. However, his only experiences of flying transport aeroplanes were the military types at Boscombe Down, so initially he had a couple of routine C of A Renewal flight tests on a BAC 1-11 with John Carrodus looking over his shoulder. It seems these went very well indeed, and he was soon working just like the other test pilots, albeit with the benefit of Keith Perrin's guidance as his flight test engineer.

Nick also needed to gain experience of the small twin-engine propeller aeroplanes. This was when I began flying with Nick. By coincidence, our first flight together was a C of A Renewal air test on a Rockwell 685F, similar to my very first CAA flight test with Gordon Corps. This had the same nose-wheel steering system, with directional control on the runway accomplished by application of the wheel-brake pads on the pedals. This was new to Nick, but he took to it like a duck to water.

This Rockwell 685 had a new modification in its elevator control linkage, which the CAA had not previously evaluated and which needed to be assessed during this flight. For maximum flexibility of the location of occupants within the cabin, it was desirable to have as wide a range as possible of the approved centre-of-gravity (c.g.) location. At the aft c.g. limit, the control forces often became very light and the critical case of stick force per g (the control force required to execute a nose-up or nose-down manoeuvre) was so low that it only just met the minimum requirement in the BCAR or FAR regulations (see Chapter 4). The flight crew was always expected to check the weight and balance calculations to ensure that occupants were located in positions that maintained the aeroplane within the c.g. limits. However, sometimes, with a lot of baggage and a full cabin, this may not always be possible to achieve.

This was the case with the Rockwell 685F. The FAA had approved a bob-weight modification which increased the stick force per g and allowed an extension to the aft c.g. limit. As described in Chapter 4, this was a fairly common 'fix' for this problem. This bob-weight was incorporated into the elevator control linkage in such a way that, when subjected to the manoeuvring g-force, it acted against the pilot's input, thereby increasing the stick force he needed to apply. In all other situations, it had little or no effect on the normal pilot's control forces. Nick duly checked the stick force per g at the new extended aft c.g. limit, using his newly-acquired ubiquitous CAA spring balance, and we confirmed it to be satisfactory.

Apart from his continuing association with small twin-engine aeroplanes, Nick was the Flight Department's BAC 1-11 pilot of choice and soon expanded his expertise to many other types. I remember one occasion when I was with him on his first flight in a Boeing 707. There were still very many of these aeroplanes in service all over the world with a variety of airlines; this one was a British Airways C of A Renewal air test. It was a particularly turbulent day and the British Airways pilot in the right-hand seat was a little apprehensive. On all such flight tests, the CAA pilot would do all the flying from the left-hand seat normally occupied by the captain, but the company pilot in the right-hand seat would be the legal captain. Nick got airborne without any difficulty and immaculately accomplished all the scheduled tests. The Boeing 707 was notoriously difficult to land in a cross-wind and that day it was close to the specified limit. Nick was obviously on a steep learning curve, but he controlled the approach and landing speeds well and had just the right amount of bank angle to maintain the runway heading; not an easy task as the outer engine could get quite close to the ground at touch-down. I know the British Airways pilot was impressed.

In all my flying with Nick, he was the most accomplished of all the pilots at achieving and accurately maintaining precise test conditions. Being an ex-single-seat fast-jet pilot, he often liked to pause after each test and make his own notes.

This could have been a bit of an irritant for a seasoned flight test engineer who was supposed to be responsible for recording all the results as well as defining the test conditions, but I appreciated his past experience and found that his understanding of the results was second to none. Over the months, Nick came to appreciate the teamwork situation he had with the flight test engineers and he reserved such note-making to the most critical testing.

One particular highlight for me during this early period was three flights in Concorde with Gordon Corps. Keith Perrin, as chief flight test engineer, rightfully claimed the responsibility for the Concorde project, but with his other workloads, there were occasions when he was away. Gordon felt he needed a nominated back-up for Keith and, very encouragingly for me, he agreed with Dave Davies that I should be that person. It was still a year before Concorde would enter service with British Airways and Air France. The manufacturers (British Aerospace and Aérospatiale in France) still had a lot of development testing to be done, and Gordon often needed to make his assessment of any potential change to the flying qualities or emergency procedures.

I soon appreciated, with Gordon's guidance, that planning a flight-test programme for Concorde was significantly more involved than for a normal subsonic aeroplane. The flight conditions necessary for each individual test had to

Concorde prototype landing, nose and visor down. (*Airliners.net, Steve Fitzgerald*)

be planned in sequence following the standard operational flight profile: take-off, initial climb, the acceleration climb phase through Mach 1, up to Mach 2 at about 50,000ft and then on up to the maximum altitude of 65,000ft, followed by the deceleration and descent, low-speed handling, and finally the landing. If you missed a test or needed to repeat anything, a quick decision was necessary; otherwise a lot of fuel and time would be wasted in having to regain the required altitude or Mach number. The low-speed handling tests, including stall approaches, were the last items on the programme before landing. Technically, a slender delta wing does not 'stall', but it enters a mode at a high angle of incidence where the lift is entirely derived from the suction of vortexes that are created from the highly-swept wing leading edge. However, there is a minimum acceptable speed at which the elevator and aileron controls on the wing trailing-edge remain effective. This is defined as V_{MIN}, which for a conventional aeroplane would be the stalling speed. To prevent the airspeed reducing below this V_{MIN}, a 'stick-wobbler' is installed, which acts to apply a nose-down elevator input. This works rather like the stick-pushers fitted to 'deep-stall' aeroplanes but, on the Concorde, there was a reluctance to call it a 'pusher', a name that carried a bit of a stigma with the French, probably as it was deemed to have been a British invention.

A great deal has been written about Concorde that would more than fill a book, but there are some aspects I found particularly memorable. The aerodynamic

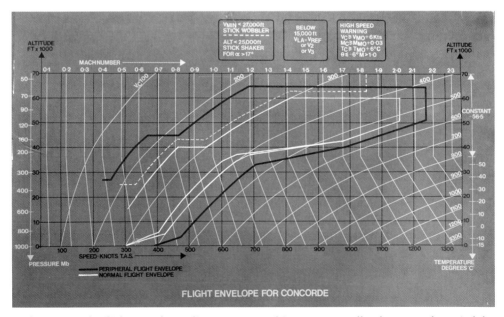

Early Concorde flight envelope diagram; note this was manually drawn and copied by photograph. (*Civil Aviation Authority*)

characteristics of the ogival-shaped slender delta wing were such that the centre of lift moved significantly aft with increasing Mach numbers. It was necessary, therefore, to move the centre of gravity aft as the Mach number increased to maintain acceptable control and stability characteristics. This was accomplished by transferring fuel to a tank in the rear fuselage. The weight and c.g. were continuously computed from the contents of all the twelve separate tanks in the wing and the tail tank. The computed c.g. position was displayed on a unique gauge on the pilot's instrument panel. However, it was not enough for the pilot to know the c.g. position; he needed to know whether or not it was in a safe position for the Mach number at which the aeroplane was flying. The innovative solution was to incorporate a pair of pointers on the c.g. indicator, which moved in accordance with the actual Mach number, to show the acceptable range of c.g. position for the current Mach number. The pilot could immediately see that the c.g. was in the appropriate location. Similarly, the Mach meter had maximum and minimum limit pointers, driven from the c.g. position computer, to show that the indicated Mach number was within the acceptable range for the actual c.g. position.

The second interesting aspect was the means of controlling the airflow into and out of the engine using moveable ramps in the intake and nozzles in the exhaust section. The Rolls-Royce Olympus engines were basically the same as those installed in the Avro Vulcan bombers. Being a subsonic engine, the airflow into the engine needed to be controlled so it would be subsonic. The secret was a series of ramps in the intake ducting that managed the shock-waves, ensuring the air entering the engine was effectively subsonic. Nozzles in the exhaust section were also necessary. Each engine had its own computer to control the movement of these ramps and nozzles, taking data inputs from every conceivable influential aspect of the aircraft and engine including attitude, incidence, sideslip, airspeed, altitude, air temperature, weight and many engine parameters. Very fortuitously, the advent of digital computers came to the rescue at this critical moment in Concorde's development. The digital computing technology of the time meant that each of the computers was the size of an attaché case. It is sobering to realize that if, at the time of the early design work, the designers had realized how much computing power would be needed to successfully control the ramps, the project may never have gone ahead. With the analogue computing technology at that time, such a computer would have had to take up a significant part of the fuselage!

One other feature that sets Concorde aside from other civil aeroplanes and is externally more obvious is its ability to droop the nose to allow the pilot a reasonable forward view during take-off and landing. Because of its slender delta (ogival-shaped) wing, at low speeds it obtains most of its lift from the suction of the vortex generated above the wing. In a conventional wing, lift is generated from the differential pressure caused by the different speed of the air flowing above and below the wing; at higher speeds the slender delta wing behaves in the same way. To generate the vortex, the aeroplane has to fly in a very nose-high attitude. It was obvious to the designers of Concorde that the slender nose extending well ahead of the cockpit would obscure the pilot's view to an unacceptable degree, hence the nose was designed to be lowered for take-off and landing. At the same time a visor with transparent panels in front of the windscreen was added which could also be lowered to improve the pilot's forward view. When on approach to land, the nose is drooped down by 12.5 degrees to allow the pilot a reasonable view of the runway. On take-off, so that the pilot can see where he is going, the nose only needs to be drooped by 5 degrees. It then stays at 5 degrees until the airspeed reaches 250 knots when it is fully retracted, unless an immediate landing is necessary, when it will be drooped to 12.5 degrees. Whenever the nose is drooped, the visor will also be down.

Our 'working lunch' pub sessions were never wasted, generally sharing experiences and discussing current certification problems. However, on some occasions they were the perfect environment in which to have a focused discussion on the development of airworthiness requirements to cope with new aeroplane design trends and unusual characteristics.

On one memorable occasion, John Carrodus, Keith Perrin, Don Burns and I were in a back room of the Skimmington Castle pub near Reigate. We were pondering a new developing situation that was arising from the trend to install ever more powerful engines into smaller twin-engine aeroplanes designed primarily for air-taxi and corporate purposes. The main manufacturers of such types – Piper, Cessna and Beech – were all taking advantage of the newly-available small turbo-propeller engines to replace the piston engines currently used in such types. Extra power was always a design goal and these engines, being lighter in weight than their piston predecessors, provided a double advantage. These factors had two consequences. Firstly, being lighter, the engines had to be mounted further forward of the wing in order to keep the centre of gravity in the right place. Secondly, the extra power often necessitated a larger diameter propeller. The combined effect of these factors significantly increased the amount of air flowing over the wing behind the propeller. This was great for increasing aerodynamic efficiency but, in the event of one engine failing, the loss of this airflow on one side inevitably resulted in a much greater difference between the lift and drag of the two wings. This was in addition to the obvious directional imbalance due to loss of thrust on one side.

The direct effect of the asymmetric thrust (the yawing of the aircraft nose towards the side of the failed engine) could be calculated with reasonable confidence. The size of the fin and rudder were increased accordingly to meet the requirements to maintain directional control at the low speeds associated with take-off and landing.

The rolling tendency caused by the loss of wing lift behind the failed engine had been of little concern with the previous piston engines. Specific requirements for adequate roll control in other flight conditions had been invariably more critical in determining the size of ailerons (wing controls for rolling the aircraft) than the engine-failure roll control case.

Some turbo-propeller versions of the above-mentioned types were beginning to give concern in this respect. The 'crunch' came with the CAA investigation for potential UK certification of the Mitsubishi MU-2. This aeroplane, originally of Japanese design, was being manufactured under licence in the USA and certificated under the jurisdiction of the US Federal Aviation Administration (FAA). The MU-2 is a corporate/business turbo-propeller aeroplane with a relatively short wing, mounted on top of the fuselage. Hence, the propeller diameter is a very large percentage of the wingspan and consequently, the effect of an engine failure on the differential airflow over the wing was very significant.

Mitsubishi MU-2, the type that led the CAA to develop new flight controllability requirements. (*ABPic, K. West*)

John Carrodus was the CAA test pilot for the investigation and Don Burns was the flight test engineer. One of the standard flight tests is to fail an engine at the critical take-off speed. For obvious safety reasons, this was normally done at a height of at least 1,000ft above the ground. Before attempting this dynamic test, it was normal practice to check the controllability, with one engine shut down and the other at full power, by slowing down carefully from a speed comfortably above the critical minimum control speed declared by the manufacturer. In the case of the MU-2, when this speed was reached, it was only just possible to maintain straight and level flight with full rudder and full roll control. Academically, the existing specific requirement was satisfied. However, during the subsequent dynamic test of failing the engine at the critical take-off speed, it was just possible to maintain directional control with the rudder, but even with a brisk application of full control-wheel, the aircraft continued to roll uncontrollably and the airspeed had to be increased by easing the nose down to regain control. On the MU-2, the problem was compounded because the roll control was achieved by the use of a spoiler on the upper surface of each wing rather than conventional ailerons. Without having ailerons taking up space on the wing trailing-edge, this arrangement allowed the whole of the (short) wing span to be fitted with high-lift flaps for take-off and landing. To roll the aeroplane to the right, the spoiler on the right wing would extend, thus reducing the wing-lift on that side and effecting a roll in that direction.

Although the FAA had granted the MU-2 their Type Certification (see Appendix II), John Carrodus and Don Burns had no doubt that this was unacceptable, and they were forced to cite a general ('catch-all') requirement in the BCAR to cover the case: 'The aeroplane shall be safely controllable…under any normally expected operating conditions or in the event of sudden failure of any engine.'

Back home at Redhill, John and Don had no difficulty in convincing Dave Davies and the others in the Flight Department that this was the correct decision. However, it was also important that this situation should be addressed by developing a new specific requirement. Hence, in the relative quiet of the back room in the pub, thoughts and ideas were shared. The solution that evolved was simple. A requirement should be introduced that specified a minimum rate of roll (degrees per second) with one engine inoperative, which had to be achievable at the certificated minimum take-off speed (V_2) and minimum landing speed (V_{AT}) as specified by the manufacturer.

Any new CAA requirement, or change to an existing requirement, was first published as a BCAR Paper that could be included as part of the certification requirements for any new type. Such a Paper was drafted for this new requirement;

it was usefully applied in several Type Certification evaluations and was eventually fully adopted into the BCAR for both small (Section 'K') and large (Section 'D') aeroplanes. Ultimately, the FAA also introduced a similar requirement into their FAR and, much later, it became part of the European Joint Airworthiness Requirements (JAR). (See Appendix I.)

Largely because of this characteristic, the CAA was never able to grant a Type Certificate to the MU-2. Although the FAA maintained their Type Certification, after a number of relevant accidents (some of them fatal), the FAA carried out a special safety evaluation. This concluded that, although the aeroplane met all their specified safety requirements, the FAA determined that some type-specific special pilot training was necessary. This was mandated using a Special Federal Air Regulation (SFAR) directed at all current MU-2 pilots. After this, the accident rate noticeably improved, but accidents following engine failures did still occur and at least one was fatal.

With the rapidly-changing capability and complexity of modern aircraft designs, the adequacy of the airworthiness requirements was being continually challenged. The Flight Engineering Section of the Flight Department was primarily responsible for keeping the development of the flight requirements up to date with the latest developments. This was an onerous task. From my involvement with light aeroplanes (those below 5,700kg) in my early CAA years, it was very noticeable that autopilots were becoming a standard feature for this class of aeroplanes. However, there were no specific autopilot requirements in BCAR Section 'K'; only a reference that the principles of Section 'D' should be followed. This was becoming a problem because it was open to interpretation and, sooner or later, there would be controversy; some manufacturers may find they had been required to meet a tougher interpretation than others. I discussed this with Denis Warren and John Rye who were the two members of the Engineering Section responsible for all aspects of flight systems. They strongly sympathized with my worries, but their in-trays were piled high with work on requirements for Automatic Landing, Concorde, GPWS, Flight Data Recorders and many other subjects.

I felt that, with my strong background in autopilot work, I could prepare some proposals that would be more commensurate with these lower-weight aeroplanes. I discussed it with John Carrodus who, of all the test pilots and flight test engineers, I had observed to be the one most ready to spend time working on potential requirement improvements. He was immediately supportive and encouraged me to have a go. He also took the trouble to get Keith Perrin's support. I went to talk to Denis Warren and John Rye to outline the principles I would aim for. They also gave their full support but, as it would be in their sphere of responsibility, they did ask that I kept them in the loop from time to time.

One of the concepts implicit in BCAR Section 'K' and the equivalent USA regulations in their FAR Part 23 for small aeroplanes was that, since the speeds in the take-off and landing situations were much lower than the large transport aeroplanes and the energy was much less, a crash in these circumstances was highly likely to be survivable. With this principle in mind, I set to work in any spare time I could find to draft a BCAR Paper. Obviously, my starting-point was the BCAR Section 'D' requirements. I went through each item in turn and made an initial judgement of whether to keep it unchanged, alleviate it in some way or delete it altogether. There was a standard format to follow in preparing such a Paper, which included a section explaining the reasons and justifications. This was a great help in preparing a cogent set of requirements, probably about 75 per cent of which were a logical extension of Section 'D'. The remainder, particularly some of those that I considered needed to be alleviated in some way, required much thought and judgement. With one or two gaps in it, I showed my first draft to John Carrodus, Denis Warren and John Rye. John Carrodus was in and out of the office for the next week or so, and I had to wait for his response. It went somewhere low down in the in-trays of Denis and John Rye. To my delight, John Carrodus had few criticisms and made some useful suggestions on how to fill the gaps. Spurred on, I 'bit the bullet', finished it off and took my 'final' draft to John Rye (Denis was away) with a plea to not let it go into his 'black hole' in-tray. To give him his due, he obviously found some time to go through it in the next few weeks; possibly, I suspected, under some pressure from John Carrodus. John Rye made some constructive suggestions and gave it his full support, after which it was subsequently published. It was satisfying to have been a part of the teamwork evident in the Flight Department, which crossed section boundaries whenever there was a compelling priority.

After nearly a couple of years in the CAA, during which time I had predominantly been dealing with many of the older aeroplane types and a variety of light aeroplanes, I was ready to be entrusted with some new responsibilities. I was allocated the new and developed types from Fokker, Dassault and those of Hawker Siddeley Aviation at Hatfield. The first of these that required a flight test assessment for the granting of a CAA Type Certificate was the Fokker F28-4000. The Fokker F28 was a medium-sized commuter jet with two engines tail-mounted on the rear fuselage; similar to a small BAC 1-11. Apart from the 'light-touch' certification flight assessments of new light aeroplane variants I had made with Darrol, this was my first opportunity to be involved with a serious Type Certification assessment, albeit not a totally new Type Certificate.

The Fokker aeroplanes were Type Certificated under the jurisdiction of the Dutch airworthiness authority the *Rijksluchtvaartdienst* (RLD). In accordance with the ICAO convention, any country wishing to certificate an aircraft manufactured under the jurisdiction of a foreign authority, for use on their own register, would grant its own Type Certificate as a 'Validation' of the Type Certificate granted by the authority of the State of Manufacture; in this case the RLD. (See Appendix II for further explanation.) Hence, the principles for a validation assessment start from the premise that we, the CAA, were fully able to accept the certification statements of the RLD. We would only wish to check compliance in any critical areas as well as those cases where our requirements and interpretations of the BCAR were different to the RLD requirements. In common with most foreign authorities, the RLD used the American FAR as the basis for their own certifications and we were quite familiar with the areas of difference.

The Fokker F28-4000 version had a longer fuselage, a capacity of about eighty-five passengers, an increased wing-span and up-rated Rolls-Royce Spey engines. These changes were significant enough to necessitate an assessment of the more critical aspects of the flying qualities. Apart from the obvious effect on the flying characteristics, the only other aspects of these changes which needed a CAA assessment were the structural and cabin safety implications; these were being carried out separately by the relevant CAA specialist engineers.

Fokker F28-4000 at Schiphol, Amsterdam, January 1977. (*Author*)

The F28 had its tailplane mounted on top of the fin like the BAC 1-11 and the Trident, known as a 'T' tail. The Trident was the first UK aeroplane to adopt this configuration, and the first to exhibit the 'super-stall' phenomenon. This is when the aeroplane slows down to the point where the wing stalls. As explained earlier in Chapter 4, the disturbed airflow breaks away from the upper surface of the wing and flows upwards over the tailplane. The tailplane itself effectively becomes stalled and can no longer exert the nose-down corrective force to recover from the stalled condition. The result is potentially catastrophic. The aeroplane enters a stable, uncontrollable situation where it descends quite rapidly in a slightly nose-high attitude. The solution, first developed on the Trident, was the introduction of a stick-pusher, programmed to operate just before the actual stall is reached; this recovers the aeroplane nose-down to increase the airspeed. Such a device was also installed on the BAC 1-11, the BAC VC-10 and, in the USA, the Boeing 727, and was found to be necessary on the F28. It was an essential part of the flight test programme that the manufacturer should thoroughly check the functioning of the stick-pusher in all flight conditions.

We hoped to carry out both the two flights – one at high weight, forward c.g. and one at low weight, aft c.g. – in one day. However, we needed some time to go through the Fokker flight results before formulating our detailed programme of tests, and so we took the morning flight to Amsterdam the previous day to give us the afternoon for this. From the CAA's previous experience of the original F28 certification, there were few critical aspects of the certification flying characteristics. Increasing thrust usually brings with it directional control problems with one engine inoperative. However, lengthening the fuselage moves the fin and rudder further aft which is compensatory. Increasing the wing span on a swept wing can often increase the tendency for a sudden wing drop at the stall but, on the positive side, it should reduce the stalling speed slightly. Clearly, it would need a good look at the effectiveness of the revised stick-pusher and its operating speeds. I had read through the previous CAA Flight Test Report and after a thorough review of the Fokker flight test results combined with Gordon's previous knowledge, we were quickly able to put together a programme for each flight. We always included a five-minute climb performance check with one engine inoperative at the beginning of the high-weight forward c.g. flight, particularly as the manufacturer will always wish to claim maximum credit for increased thrust and more wing area. Additionally, on new variants, there were usually some changes to the flight-deck instruments and controls that needed to be assessed and approved.

We made an early start and took off on our first flight at 09:25 with the aeroplane loaded close to maximum take-off weight and at the forward c.g. limit. Henk

Themen was the Fokker test pilot in the right-hand seat with Gordon flying the aeroplane from the left seat. As well as the climb performance and stall speed checks in a variety of configurations, we went through the critical controllability cases. After lunch, we followed with the aft c.g. flight, during which we found a few marginal characteristics. By then it was late afternoon and we needed to review the instrumentation records before reaching our final decisions. Fokker utilized the same type of photographic trace recorders that were used at Woodford, and we had to wait until next morning before they were developed. Back at the hotel that evening, as was customary, we prepared our debrief conclusions, as far as we could, ready for the following morning.

The CAA Flight Department had devised a system of classifying their findings into four categories:

1) 'Snags' which were unacceptable issues, where a modification or change to the operating limits would be required.
2) 'Major Criticisms' which were characteristics or features considered to only just meet the requirements.
3) 'Criticisms' where there was no issue of non-compliance but where it was considered that the aeroplane would be improved if changes were made to address the criticism.
4) 'Unresolved Items.' These were areas where either further testing was required, or the CAA flight test team felt that they needed to take the issue back to the CAA for further consideration.

Each item noted in flight was allocated into one of these categories for the debrief. Clearly, most features were totally acceptable and did not get a mention in the debrief, but Gordon took the trouble to be very complimentary on many of the flying characteristics and flight-deck features.

The only problem of note was the wing flap selector. It was too easy to inadvertently move the selector beyond the next flap position. If the flaps were inadvertently lowered too far, the airspeed could be greater than the maximum speed limit for that flap angle. Similarly, raising flaps too far at a low airspeed could lead to the wing entering a stall condition. The solution was to require a modification to the flap selector lever which introduced a 'gate' at each flap position, requiring a deliberate movement of the lever through the gate to the next setting.

Gordon was sometimes particularly keen to make his own assessment of certain engineering aspects of an aeroplane that could give cause for concern. Prior to this visit, there had been a few cases of baggage doors inadvertently opening in flight without warning on some aeroplane types. On transport aeroplanes, it was normal for the under-floor baggage hold to be within the pressurized fuselage, so any failure of the baggage doors to stay latched would result in a loss of cabin pressurization. The design of the latching system on some baggage doors was not to as high an integrity as the main passenger cabin doors. Although they had the same safety features, the mechanism could be susceptible to damage from baggage or the baggage-handling trolley impacting on the door during loading. If unnoticed it could be possible to close the door with the normal 'locked' indication showing but with the latches not fully engaged. When walking through the hangar before our flying, Gordon made a bee-line to an F28 that was undergoing maintenance and asked if he could open a baggage door. I immediately realized what he was up to and we both assessed the opening mechanism and the locking indication system together. It was a well-engineered design with none of the frailties present on the doors of other aeroplane types that had failed. This was another broadening of horizons that I took on board. Although this particular issue was clearly the responsibility of the mechanical and pressurization systems specialists, anything where an indication or warning was presented on the flight-deck had a cross-discipline connotation.

A few weeks later, I was off to Hatfield with John Carrodus to do the certification flight assessment of the latest version of the Hawker Siddeley (British Aerospace as it was about to become) 125: the 125-700. The 125 was a medium-sized business jet with the engines mounted on the rear fuselage and with a 'T' tail configuration. Like many other aeroplanes with this configuration, when initially designed it was anticipated that the 125 would be susceptible to the 'deep-stall' characteristic. However, after great care in the early stall testing, surprisingly it was found to have a good natural nose-drop at the stall in all configurations and did not need to have a stick-pusher installed.

The 125-700 was fitted with new American Garrett TFE731 turbo-fan engines in place of the original Rolls-Royce Viper engines. By modern standards, the Vipers were very noisy and thirsty engines (like those fitted to the Viper Shackleton from my Avro days). The TFE731 had a maximum thrust of 3,720lb; this was about the same thrust as the last version of the Viper installed in the previous version, the 125-600. Our approach to assessing the Hawker Siddeley Aviation test reports, and our flight programme planning, followed a similar pattern to that of the F28-4000 with Gordon.

However, this being a UK certificated type, the CAA, as the Primary Certification Authority, would be fully responsible for establishing that all the

HS 125-700 with deep-stall recovery parachute in tail; John Carrodus by the nose, at Hatfield, March 1977. (*Author*)

flying qualities met the appropriate requirements of BCAR Section 'D' so that they were able to issue the UK Type Certificate. Hawker Siddeley Aviation (HSA) was a CAA 'approved company'; fully accepted as being competent to carry out all aspects of design and testing necessary to show compliance with the specified CAA requirements. Hence there was no necessity for the CAA to double-check every aspect of the compliance testing and, in a similar way to our approach to validation testing, we would only need to check the more critical areas. (Again, see Appendix II.)

Whenever significant changes were introduced to the 125 design, HSA sensibly explored the stall-handling characteristics with caution in case the changes had 'tipped' the aerodynamic characteristics into the 'deep-stall' regime. As a precaution, they fitted a parachute in the tail that would be deployed in the event of a 'deep-stall' to introduce a nose-down pitching moment. Once the nose was pointing downwards, full power would be applied and the parachute jettisoned so that the airspeed would increase. Although HSA had thoroughly checked the stalls in all configurations and found no problems, the tail parachute was retained for our stalling tests as a precaution, just in case there was anything different in our technique or the test conditions that may have had an unexpected effect.

The introduction of the TFE731 engines came with a digital engine control system and the latest design of engine instruments. At the same time, HSA had taken the opportunity to fully upgrade all the flight-deck instruments, warnings, switches and system indications to the latest standards. There was a lot to assess here and we had to include a flight at night to check that all this gave satisfactory indications in the dark.

As usual, John flew from the left-hand seat with the HSA Test Pilot Mike Goodfellow in the right seat. The very experienced de Havilland/HSA flight test engineer Roger de Mercado was on board to run the instrumentation, keep his eye on our results and write copious notes. We began with the high weight, forward c.g. flight which comprised mostly climb performance, stalling speed checks, and the more critical controllability cases. This took nearly four hours of concentrated flying. Operating from Hatfield, just north of the London Air Traffic Control Zone, it was often challenging to find enough uncontrolled airspace (which includes all the airways to and from Heathrow) in which to do the testing. The next day we took off expecting to get stuck into the aft c.g. flights, but we were soon thwarted by the failure of one of the engine control computers, so it was back to Hatfield. It was an opportunity to assess the emergency procedures developed for engine control in such circumstances; we had this on the programme for our final flight. We consulted the proposed Emergency Procedures to use for this situation. Caution was necessary when making large throttle movements and close monitoring of the engine instruments was necessary to ensure engine limits were not exceeded. It worked OK and the level of additional workload was judged to be acceptable for such a failure situation. However, we decided that the Flight Manual procedures needed to be a little more explicit.

It was another week before we could schedule the remainder of our flying which would include the low-speed minimum control speed and stall-handling tests. This took two flights as the minimum control speeds had to be checked at as low a weight as possible, and without the tail parachute installation that could have affected the directional control and stability. Everything went well, and we were left with only a few points to raise in the debriefing the following day. The HSA, Hatfield, pilots and engineers were well calibrated to the requirements and interpretations of the CAA Flight Department and we found no significant issues.

In early 1977, shortly after the 125-700 flight-test programme, John Carrodus dropped a bombshell into the settled state of the Flight Department. He had been offered and accepted the post of Operations Director with Cathay Pacific Airways in Hong Kong. The double whammy was that Don Burns had also been recruited by Cathay at the same time to be their chief performance engineer.

At first, my reaction was disappointment; I had been flying a lot with John and I felt we had a good rapport and working relationship. However, clearly much decision-making was required to re-allocate the current aircraft type responsibilities of John and Don, as well as the forthcoming new certification projects, to the other test pilots and flight test engineers. Behind closed doors with advice from John, Don and Gordon Corps, it was the responsibility of Dave Davies and Keith Perrin to work on a new plan. I do not remember giving much thought to the likely outcome, but the result certainly surpassed any expectations I might have had. Of Don Burns' aircraft responsibilities, the major prize for me was the Lockheed L-1011 TriStar, which at that time was in the middle of major developments. Additionally, the Dassault Falcon business jets, the Rockwell and Piper twin-turbo-propeller aeroplanes and, in the UK, my old friend the Avro 748 came my way. It was comforting to realize how much responsibility was being entrusted to me and that, in particular, although only a few years out of Hawker Siddeley Aviation, I was being relied on to be impartial in future 748 certification issues. By chance, almost immediately, I was off back to Woodford with John Carrodus to evaluate the handling and performance characteristics of a 748 fitted with a large radome in the nose. This was an HSA 'private venture' to offer a cheaper alternative maritime patrol aeroplane for smaller countries who could not justify the expense of a Nimrod. It had to be certificated to civil (CAA) standards as many countries had their 'coast-guard' aircraft on their Civil Register. The other major bonus for me was to be allocated the certification flight-testing of the Bandeirante, a new Brazilian aircraft manufactured by Embraer.

Along with my new allocations, John's test pilot duties were also re-apportioned. Again, the big winner was the other relatively 'new boy', Nick Warner. Nick and I had flown quite a lot together since he joined in early 1976, but mostly on routine flight-testing for renewal of Certificates of Airworthiness on BAC 111s, Boeing 707s and various small twin-turbo-propeller types. We were now to be entrusted as a team to take on the UK certification flight-test responsibilities for two new major projects: the Lockheed TriStar L-1011-500 and Embraer EMB 110 Bandeirante. This was not going to be straightforward.

The TriStar was a state-of-the-art modern large transport aeroplane and the latest version, the L-1011-500, had some unique design features. In order to get up to speed with all this, it was proposed that Nick and I should spend three weeks at the Lockheed plant in Palmdale, California on a specially-tailored course to cover in detail all the design features and engineering systems. For both of us, this turned out to be a most valuable opportunity.

The Embraer EMB 110 Bandeirante was a different situation. So far, no-one from the UK CAA had been involved with any aircraft from Embraer, nor had anyone had experience of the Brazilian Aviation Authority. It was deemed by Dave Davies (rightly) that this was not a situation where it was sensible to send the least experienced flight test team. The neat solution was for me to team up with John Carrodus for the first visit and Nick Warner would take over all subsequent test-piloting. A fortunate circumstance of timing meant that John could just make this visit before his departure from the CAA.

Life for me could only get more interesting, and it did.

Chapter 10

New Experiences One After Another

So there I was in May 1977, boarding an Air France Concorde at Paris Charles de Gaulle airport for the flight to Rio de Janeiro, with John Carrodus and the rest of the CAA team. Beyond my wildest dreams; how did it happen? Denis Murrin was head of the CAA Foreign Aircraft Certifications. His primary role was to manage the teams of CAA specialists who would visit the manufacturers to undertake the detailed investigation of the aircraft type in question. He was a larger-than-life Churchillian character, respected throughout the world for the way in which he would tightly and objectively manage these investigations.

Once a manufacturer had made their Type Certification application to the CAA, Denis would give the manufacturer an estimate of the investigation costs. As mentioned earlier, all the CAA's costs were directly financed, as appropriate, by the manufacturers, airlines and owners. In general, this was from the fees they paid for the Certificates of Airworthiness issued to each individual aircraft on the UK Register and from Operating Certificates issued to airlines and other operating companies. However, in the case of the initial design investigations for Type Certification, the man-hours, travel, accommodation and subsistence costs were charged directly to the applicant. Denis, always one for spotting an opportunity, had noticed that the flights to Rio did not arrive until late in the evening with no possibility of catching a flight that same day to São José dos Campos where Embraer had their factory. A night stop in Rio would be necessary and by the time the team had made the flight the next day and checked into the hotel, a good deal of that day would be gone. At that time, Air France was operating Concorde on the route from Paris to Rio. Because of the much-reduced flight time, it arrived in the early afternoon, leaving sufficient time to make the journey to Embraer the same day. Denis had reviewed the cost differential between first-class and Concorde and determined that, per person, it was a little less than the total cost of an extra day on the programme. He had put this case to Embraer and they accepted it.

These CAA certification investigating teams usually numbered between seven to ten people depending on the complexity of the aircraft being investigated. As well as the team leader, there was the flight test team (usually test pilot, flight test engineer and performance engineer), and engineer specialists in the fields of structures and systems (flying controls, mechanical, hydraulic, electrical, engine installation/fuel systems, cabin safety/environmental control and avionic/flight control). In the case of the smaller aeroplanes with mechanical flying controls and simpler systems, it was usual to have one systems specialist cover more than one discipline and the team leader would often be able to deal with the cabin safety issues and sometimes the mechanical systems, thus reducing the team size. As well as John Carrodus, myself and Andrew McClymont, in this team we had Peter Richards (structures), Bob Casbard (flying controls, mechanical and environmental control systems), Brian Perry (electrical and avionics systems) and Vin Wills (engine installation and fuel systems). Denis dealt with the cabin safety issues. This was the first time that the CAA had any experience of products from Brazil and we had no first-hand knowledge of the certification standards of their controlling Airworthiness Authority, the Ciência e Tecnologia Aeroespacial (CTA). So, for this visit, on the team we also had Colin Turner, a CAA maintenance and manufacturing surveyor, whose task, along with Denis, would be to review the capabilities of the CTA. This gave a total team size of nine.

The subject of this evaluation visit was a twin-turbo-propeller, eighteen-seat commuter aeroplane: the Embraer EMB 110 Bandeirante. We were to assess two different versions: the 110P and 110E. They were aerodynamically identical and had the same Pratt & Whitney PT6A-27 engines giving 680 shp (shaft horsepower); the E version simply identifying that it was configured as an executive rather than normal passenger transport.

The normal direct flight to Rio took about thirteen hours. Concorde did not have the range to do the journey non-stop, so there was a scheduled refuelling stop at Dakar in West Africa. Nevertheless, two legs of about three and a half hours plus forty-five minutes in Dakar made it an easy, relaxed journey. Flying at twice the speed of sound, being able to see the curvature of the earth and enjoying the exemplary in-flight service was for me a memorable experience.

However, the stop-over at Dakar was in itself memorable for several reasons. The remote airfield had been specially adapted to service Concorde and its passengers. There was a pleasant transit lounge with an extensive, very cheap duty-free shop. Here, Denis began to exercise his authority on the team members! He insisted that each of us should purchase one bottle of good-quality Scotch whisky. When I protested that I did not drink Scotch and would prefer to buy a bottle of gin, the other team members, having much experience of working with Denis, suggested that I should comply without question. I had heard that Denis was partial to a dram or two in the evenings, but I did not conceive that he would be able to consume this amount in less than two weeks. Nevertheless, I duly purchased my bottle of

Air France Concorde in Dakar en route to Rio de Janeiro, May 1977. (*Author*)

Scotch together with a bottle of gin for my own consumption. Two bottles were the duty-free allowance into Brazil.

On our arrival in Rio, we were expecting to be transferred to a flight to São José dos Campos. However, after passing through customs, we were met by a contingent of Embraer executives including the Engineering Director Guido Pessotti. They promptly whisked us into a fleet of cars and took us to the five-star Meridian hotel on Copacabana beach. They had laid on a special reception for us which included a dinner and show. So much for Denis's cunning plan of saving a day! He did protest strongly, but they were determined to demonstrate the renowned Brazilian hospitality.

So, somewhat a little the worse for wear, late the following morning we were driven back to the airport to be flown the one-hour flight to Embraer's airfield in one of their corporate Bandeirante aircraft. We arrived in time for lunch which included aperitifs and wine! In mid-afternoon Denis was finally able to commence his presentation of the evaluation objectives and programme, followed by the introductions of our team and Embraer's lead engineers with whom we would liaise. Effectively, we were then a day behind Denis's carefully planned programme, so some rescheduling had to be hastily worked out.

Without exception, the Embraer engineers were a highly-motivated and capable team of people. Their ability to understand, speak and read English was impressive. The language had been one of our concerns, but the communication and mutual understanding was so good that we were able to make up much of the lost time. We were treated to an extensive lunch every day in the executive dining room. We declined the wine, but the waiters were always ready to pour coffee from their jugs. This was traditionally very strong and made from Brazilian beans but, being so strong, sugar was always added in the jugs. I was the only one who could not drink sweet coffee, but after they realized this, they always had one coffee jug with no sugar just for me. For John Carrodus and me, it was quite a relief to resort to the alternative of chicken legs and sandwiches whenever our flying coincided with lunchtime.

In their irrepressible desire to show their hospitality, they wanted to invite us all out for dinner nearly every night. Denis tactfully persuaded them that this was not necessary and was not helpful to the process of the investigation. The early evenings before dinner were normally taken up with general CAA team discussions going over issues raised during the day, as well as the preparation of each team member's investigation report and conclusions. After a couple of the Embraer 'entertainment' evenings, we were able to get back to Denis's normal working process.

It was my first exposure to this concurrent team-working system. However, John Carrodus was very used to the process and had worked with Denis on many projects. Following the briefing and introductory meeting, John and I took up our temporary daytime home in the Flight Center. Embraer had recent experience of a similar flight-test assessment made by the FAA from the USA, so they were quite well prepared for our test programme requirements. The Embraer chief test pilot was a very jovial character known as Cabral. All Brazilians seemed to have a string of names and it was customary for them to select one of the names which was used both for formal and informal situations.

As stated earlier, a significant aspect of this visit required us to reach a conclusion on the competence of the Brazilian Airworthiness Authority (the CTA), to oversee the design, manufacturing and certification standards of Brazilian aircraft. The CAA had not dealt with a Brazilian-designed aircraft before. We needed to be sufficiently satisfied with the CTA's competence so that we could be in a position to issue a CAA Type Certificate as a validation of the CTA Type Certificate. The international protocol, as defined by the International Civil Aviation Organization (ICAO), required that each nation wishing to issue their own Certificates of Airworthiness to individual aircraft designed and manufactured in a foreign country had to issue their own Type Certificate as a validation of the foreign authority's Type Certificate. Where the 'Validation Authority' had individual regulations and interpretations that differed from those of the Certification Authority, they were entitled to make their own assessments. Ultimately it was the Certification Authority that would submit their statement that the aircraft, as defined, complied with the Validation Authority's regulations. Hence, it was important for the Validation Authority (in this case the UK CAA) to be satisfied of the competence of the Certification Authority (in this case the Brazilian CTA) to make this statement. (This process is described in more detail in Appendix II.) As this confidence also had to include their manufacturing control, in addition to the design, we had Colin Turner in the team. Colin was a very experienced CAA surveyor from the CAA Hatfield office where he oversaw the HSA (originally de Havilland) manufacturing site. Colin, together with Denis Murrin, carried out an audit of the CTA organization and their responsible personnel. Although many of the key CTA personnel had been recruited from senior military ranks and had no background experience of civil aircraft certification, they were found to be very knowledgeable and competent. The CTA had looked closely at the organization of the US FAA and had successfully mirrored that system, albeit on a smaller scale.

In order to finalize our flight test programme, we needed to go over the Embraer flight test reports to determine where the critical areas were. Almost all turbo-propeller-powered aircraft have the same difficult control and stability areas. As discussed before, the engine-out directional control at low speed is always critical as it limits the take-off and landing speeds. The minimum control speed has to be determined. This is the lowest speed at which it is still possible to maintain directional control with full rudder. The manufacturer is always trying to push these speeds as low as possible as the scheduled take-off and landing speeds can be no less than a fixed factor above the appropriate minimum control speed. The lower the take-off speed, the shorter the take-off distance; potentially a commercial advantage to be gained over the competition. Other critical areas are associated with longitudinal stability, generally at low speeds, where the aircraft must have a natural tendency to recover (nose-up and nose-down) from a disturbance in the flight path. In both situations, the critical cases are with the c.g. at the aft end of the certificated range. At the forward c.g. limit, it is the control forces the pilot has to exert on the control column that can become excessively heavy when manoeuvring or where there are changes of configuration (wing-flap extension/retraction or landing-gear deployment). The aircraft weight is also a factor. Other areas of particular significance are the stall handling, stalling speeds, the engine-out rate of climb performance measurements and autopilot fault simulations.

Embraer EMB 110E Bandeirante, São José dos Campos, May 1977. (*Author*)

Hence, the flight-test programme we developed required three different loading configurations: maximum weight/forward c.g., high weight/aft c.g. and low weight/aft c.g. The amount of test points at each configuration would require more than one flight in some cases, in particular the low weight/aft c.g. flight.

One difficulty that soon emerged was the lack of any seat near the flight-deck where I could conduct and observe the flight tests. Being used to larger aeroplanes with extra seats on the flight-deck or light aeroplanes with just Darrol Stinton and myself, these small commuter types introduced this new (to me) difficulty. Normally, for the company flight tests, only one test pilot was required and the company flight test engineer could occupy the co-pilot seat. In our case, although John Carrodus took the left-hand seat, the company test pilot Cabral had to be the legal commander of the aircraft and took the right-hand seat. There was a pair of seats at the forward end of the cabin which I tried for the first flight. Even with an intercom to communicate with John, I found myself, for most of the flight, standing immediately behind the pilots' seats at the back of the flight-deck compartment where I could directly read the instruments and do any timing of test results. It was also necessary to keep a record of the aircraft weight and centre of gravity position by taking regular readings of the fuel tank contents. For this first flight, comprising relatively undynamic forward c.g. tests, we managed reasonably well with this arrangement. However, it was very clear that for much of the flight testing to come, this option was not going to work. Trying to hold on to the clipboard, write down results, and use the calculator and stop-watch while standing up would not be possible with the aircraft performing dynamic manoeuvres. I had to be sitting down with the board on my knee to give me half a chance.

Before the second flight, I considered the options. I needed something to sit on just behind the pilots where I could wedge myself in such a way that I could function throughout all the required manoeuvres such as rate of roll measurements and stall handling tests. Looking around the flight office and hangar, I found just the thing: a portable wooden wheel-chock. It was triangular in shape and when turned on its side was just about big enough for my bottom, tall enough to sit on and it would fit in the available space. Clearly not having a seat belt was a potential safety issue, but I could use the cabin seats for take-off and landing. In this way, we successfully accomplished all the testing. There were times when it was extremely challenging keeping control of the seat (chock!), my clipboard and me when the Bandeirante was being thrown into rolling and pitching manoeuvres. Probably the most demanding moments were the turning stalls, starting in a 30 degree banked turn, pulling back on the control column to reduce the airspeed until the wing stalled. A satisfactory outcome was a natural recovery where the nose would go down by itself to safely regain airspeed and with the wing dropping by no more than 30 degrees of bank. Obviously, starting with 30 degrees, the bank angle could easily reach 60 degrees, and this was combined with the high nose-up attitude at the stall. With a lot of rapidly-changing data to note down, I certainly felt I was earning my pay. In such dynamic tests, the pilot and test engineer would share the reading of relevant instruments during the critical phases and the test engineer would note down the pilot's data immediately after the test was completed.

Over the next few days we completed the test programme in five flights with a total of almost seven flying hours. Partway through our programme of flights, I was offered the chance of a back-seat ride in the Embraer Xavante jet training aircraft that was being used as a 'chase aeroplane' to observe the stalling characteristics of the Embraer development aircraft. An opportunity not to be missed and at

Embraer Bandeirante 110E and 110P taken from Xavante chase-plane. (*Author*)

the end of the programme, I was subjected to an aerobatic demonstration and an opportunity to take the controls myself; a great experience.

At the end of each day, before going for dinner, the whole CAA team would congregate in one of the hotel rooms (usually the team leader's; Denis Murrin in this case) to go over all the issues found by each team member, which could be a potential problem for the certification. Inevitably, during these get-togethers a certain amount of drink was consumed, which helped to ensure that everyone would have their say! The benefit of these meetings was immense. Aircraft, being complex machines, would have many aspects of the design where more than one engineering discipline would interact. For example, the pressurization system uses 'bleed-air' from the engines, controlled and directed by mechanical and electrical systems. The flying controls were mechanically operated, supplemented with electrical systems. Each member of the team needed to be aware of how his engineering speciality interfaced with other disciplines where there may be a potential certification issue. From our flight-testing discipline, there were often many such aspects to consider. In particular, a potential failure case in the flying control systems might need a specific flight test to ensure, or otherwise, that it was manageable.

It was during the first such meeting that Denis called upon all the team to each hand over their bottle of Scotch, mentioned previously. Obviously, an explanation had to be forthcoming. I was beginning to realize that Denis was a man of many talents, opportunism being one of them! He had found out that there was a very high import duty into Brazil on foreign spirits; hence a big demand had evolved, particularly for good-quality Scotch. During the pre-visit conversations between Denis and the hotel manager, it had transpired that the manager was very keen to accept a number of bottles of quality Scotch in return for a significant reduction in the team's hotel bills. As we had purchased these bottles at a very favourable duty-free price and we were on an expense system based on a fixed daily rate, this was a win–win solution for all concerned (although definitely non-PC by today's standards!).

On the social evenings, organized by Embraer, we enhanced our working relationships with the engineers and got to know more about their personal lives. The engineering director, Guido Pessotti (known to all as Guido), was the driving force behind the rapidly-increasing competence of Embraer, both in engineering terms and the ability to find gaps in the aviation markets. He was very much in touch with the Bandeirante's design, and if any potential difficulties arose, he immediately took a direct interest.

Embraer, not satisfied with their repeated demonstrations of hospitality so far, had arranged, during the weekend in the middle of our visit, to take the whole team in one of their Bandeirantes to visit the Iguazú Falls on the Brazilian/ Argentinean/Paraguayan border. This was another unforgettable experience. At that time, I had no idea that these falls were designated one of the natural wonders of the world. They were, indeed, spectacular when viewed from the bottom. It was possible to climb up a steep path to the top of the falls. Some of us made the climb. At first sight it was not so impressive, but we noticed that there were 'guides' offering, for a small fee, to take people by rowing-boat to one of the small islands towards the middle of the falls. They seemed to be doing a regular trade taking four or five people out at a time, leaving them on the island and returning with the previous group. Three of us thought this looked like a good thing to do and, having paid, we climbed into the boat. Not far from the shore, we began to have second thoughts. The boatman had rowed out for a while, but it was soon obvious that the local currents were beginning to take control of the boat towards the falls, and what happened next was totally unexpected. The boatman stowed his oars and leaped over the side into the water! He clearly knew the depth and by keeping hold of the side of the boat, he could control it with his feet on the river bottom and guide us to the island. It was not difficult to imagine our fate if he had lost his footing and let go of the boat! We clambered onto the little island (no more than an outcrop of rock) with some relief, went as close to the edge of the falls as

Iguazu Falls. The 'outcrop of rock' can be seen sticking out at top left, just in front of the main area of the falls, May 1977. (*Author*)

View of Iguazu Falls from outcrop of rock reached by rowing-boat! (*Author*)

CAA team members (Brian Perry, forward right) anxiously waiting for boat returning intrepid team colleagues from the outcrop of rock at the top of Iguazu Falls. (*Author*)

we dared and took some amazing photographs. The procedure on the return trip was basically the same, but the boatman had to work quite hard to 'walk' the boat against the current. I seriously doubt that this 'tourist trip' is still available today!

Having recently looked at modern travel brochures on Iguazú, it seems that there is now a substantially constructed raised walkway for visitors to get to this rock outcrop, and I suspect there are considerable safety barriers erected around the rock.

On the return flight from Iguazú, it was dark and most of us were drifting into sleep. I was sitting next to Colin Turner. I noticed that he was looking intently out of the window. He realized I had taken an interest and he told me he had been watching what looked like a big forest fire. I looked out and could not see any sign of a fire. Colin pointed out of the window to what he was seeing. It then dawned on me that he was looking at the red navigation light on the port wing-tip. We had a private laugh together and settled back for a snooze.

While we were back at São José dos Campos, Embraer wished to take advantage of our presence to evaluate their single-engine crop-sprayer aeroplane, the Ipanema. This was similar in design and concept to several other crop-sprayer types, in particular the very successful Piper PA-25 Pawnee. The Ipanema had the

Embraer Ipanema crop-sprayer aeroplane with John Carrodus, São José dos Campos, May 1977. (*Author*)

usual large hopper fitted in the fuselage to contain the chemicals and extendable spray nozzles out to each wing-tip.

Because it was a single-engine piston-powered aeroplane below 6,000lb (2,730kg) in weight, it came into the category of aeroplane types that the CAA would usually accept with only a minimal investigation. However, this meant that it had to have been certificated by an Airworthiness Authority in which the CAA had confidence. Until the CAA review of the Brazilian CTA, being conducted during this visit, was completed, this situation could not be confirmed. It was agreed with Embraer that we would conduct our normal limited flight test as well as have our team look at the few specific CAA requirements appropriate for this class of aeroplane. If it turned out that we had any serious doubts about the CTA, then Embraer would reconsider their position. Before we could do anything, Embraer had to make a formal application for a UK Type Certificate. This they did immediately, and Denis Murrin was able to process their application rapidly through the CAA system by fax and phone.

Being a single-seat aeroplane, John Carrodus had to do the flying on his own, but the Embraer test pilot and I were in communication with him on the radio to give the necessary information and note down his results and comments. Obviously, he had never flown the type before and there was no opportunity for a check ride with an Embraer pilot. Furthermore, it was a tail-wheel aeroplane with a very powerful engine, which could be quite difficult to control on take-off if you were not used to it. John took it in his stride and went through the whole short test programme with no problem. Sadly for John, this was one flight test report that he would have to write himself with inputs from me rather than the other way round.

There were a few issues that arose from our flight-testing of the Bandeirante, which could be problematic to Embraer in their quest for a CAA Type Certificate. However, the version we had been flying was not the later production model for which Embraer needed the certificate. The most significant problem was a tendency for the rudder to 'overbalance' in some configurations. As increasing rudder was applied, the pilot's pedal force would begin to decrease at larger rudder angles to the point that, when the rudder pedal was released, it did not return to the neutral position; some force in the opposite direction was needed to return the rudder to neutral, an unacceptable situation. However, there was still a reasonable degree of directional stability. The longitudinal stability was very weak, but the worst case we found was in a configuration where there was no specific requirement in BCAR Section 'K'. There were also cases where the stall handling was marginal. Some future CAA flight-testing on the production version would be required to reassess these issues.

We returned home, after a very memorable visit, by the normal route: a British Caledonian Airways Boeing 707 overnight flight.

Conveniently, Embraer had planned, with their UK agents, CSE Aviation at Kidlington, Oxford, to bring the EMB 110P2 Bandeirante production version to Cranfield, for sales demonstrations the following month. This version had a stretched fuselage to incorporate a forward entrance door immediately aft of the flight-deck, enabling a potential increase in the passenger capacity from eighteen to twenty-one. The power of the Pratt & Whitney engines had also been increased from 680 shp to 750 shp. Not only was this an opportunity for the CAA to update their flight test assessment, but it was also a convenient chance to have Nick Warner familiarize himself with the aircraft and gain a working relationship with Cabral, their test pilot. They hit it off almost instantly as both had the same relaxed professionalism and sense of humour. Some design changes had been made to solve the previous issues identified on our visit. Embraer had modified the shape of the fin leading-edge to eliminate the rudder 'overbalance' and also added some small horizontal strips to the wing leading-edge; a fairly common solution to improve the nose-drop at the stall. Four flights over three days, at extremes of c.g. and weight, completed this programme and no certification issues emerged.

Embraer EMB 110P2 Bandeirante with Nick Warner, Embraer chief test pilot Cabral and the author (middle three in photo), Cranfield, June 1977. (*Author*)

Nick Warner with Embraer chief test pilot Cabral in EMB 110P2, June 1977. (*Author*)

With all these new, exciting yet demanding experiences, I felt I was walking around with a continuous smile on my face. However, another lucky twist of fate (for me) suddenly opened a new spectrum of opportunities. Keith Perrin had to go into hospital for a hernia operation. Some of his immediate flight test responsibilities had to be re-allocated. We were some way from employing a replacement for Don Burns, so it was down to Dave Cummings and me to temporarily fill the gaps.

In no time at all, full of excitement, I was travelling to Casablanca with Gordon Corps to participate in the Concorde hot-weather take-off performance testing with revised engine intake control laws. As mentioned in the previous chapter, one of the most significant aspects of Concorde's success was the way the geometry of the air intake to the engines was manoeuvred, to enable the subsonic engines to live with the supersonic airspeeds. Although Concorde had been in service with British Airways and Air France for about eighteen months, continual improvements were being explored to the intake control laws. If a worthwhile improvement was established, such as better fuel economy or reduced incidents of 'surging' at supersonic speed (the engine being momentarily starved of sufficient air), the consequences may impact on other areas. Critically, it was important that the take-off thrust or noise emissions were not compromised, hence the need to

Concorde hot-weather testing at Casablanca, August 1977. (*Author*)

repeat take-off distance and noise measurements in the critical high-temperature conditions.

British Aerospace Chief Test Pilot Brian Trubshaw was the company project pilot for these trials at Casablanca. He and Gordon Corps, being the CAA Concorde project pilot, had over several years inevitably developed a close relationship, both technically and socially. Gordon and I arrived in Casablanca the evening before the planned flights and arranged to meet with Brian for a briefing later that evening. Because British Aerospace often used Casablanca as their high–temperature test base, they had organized semi-temporary apartments for the accommodation of their trials staff. Inevitably Brian had one of the larger apartments and his wife was often able to accompany him. It must have been a very mixed blessing for her as she was usually on her own most of the day and had to cope with Brian's technical discussions and briefing with colleagues in the evenings. When Gordon and I arrived at their apartment, she immediately latched on to me as someone new and engaged me in conversation. This was a difficulty for me as I was desperately keen to participate in the discussions of the flight-test programme with Brian and Gordon, but I could not bring myself to be rude and break off from her kind attention. Fortunately, Gordon was totally understanding of the situation and filled me in on the bits I had missed during the journey back to our hotel.

The next day was both rewarding and eye-opening. Concorde was fully fuelled and loaded up to maximum weight for take-off. On the ground, a number of BAe engineers with noise-measuring equipment were positioned at specific locations on the airfield to record the noise emitted during the take-off. Unlike the earlier method of recording the take-off distance and flight path using ground cine-cameras (the Avro 748 Series 2A in Chapter 5), an inertial navigation unit, fixed in the fuselage close to the centre of gravity, was used. The aircraft was held on the brakes until the engines were at maximum power with full re-heat (afterburners: additional fuel pumped into the exhaust gases to increase thrust), and the brakes released on a countdown to start the measurements. The acceleration of Concorde never ceased to be amazing, even at this high weight. At the determined rotation speed, the control column was pulled aft, the aircraft rotated to the lift-off attitude and the take-off speed carefully maintained until the altitude of 50ft was reached. The climb was continued up to 1,000ft to complete the noise measurements. The next thing to do was to land back on the runway to refuel to maximum weight again and repeat the test. However, this was always a problem for Concorde. The maximum weight allowed for landing was limited by the strength of the undercarriage and the associated structure. In order not to add excessive structural weight for the landing strength case, it was usual for the scheduled maximum landing weight to be less than the maximum take-off weight. In the case of Concorde, this difference was about 70,000kg; very much greater than normal aeroplanes. For controlled cases with experienced test pilots, this landing weight limit could be increased somewhat, but a considerable amount of fuel still needed to be used up. For normal flying this would take an hour or so; consequently, the aircraft was fitted with a fuel jettisoning system. After each take-off, a staggering amount of about 35 tons of fuel was deposited over the nearby Moroccan desert; it vaporized before reaching the ground, but anyone on the ground would have certainly been aware of the smell.

Eight take-offs were completed in the day, mostly with a simulated engine failure at the critical take-off decision speed. Gordon and Brian shared the test flying for these take-offs. The subsequent analysis confirmed that the take-off performance and noise levels were not measurably affected by the changes to the intake ramp control laws.

The second aircraft manufacturer for which Keith was the responsible flight test engineer that temporarily fell to me due to Keith's hospitalization was McDonnell Douglas. The cargo airline, TransMeridian Air Cargo, had (almost!) purchased the first Douglas DC-8-54 to be introduced to the UK register. Although the CAA had completed the Type Certification of this 54 version of the DC-8 some years earlier, it was customary, as mentioned before, to carry out a series airworthiness flight check on the first such type as a familiarization and also to check that the critical flying qualities were as previously established.

The aircraft was due to be handed over to TransMeridian, from the previous owner, at Le Bourget (Paris) where the UK Certificate of Airworthiness would be issued, so Gordon and I needed to get there. Le Bourget was not a normal airline destination and Gordon had hired a twin-engine Piper PA-23 Aztec for the journey. Each test pilot had a training budget to keep them current on various aspects of civil aircraft operation such as flight planning and airways navigation; so, as well as being a convenient way to get there, it was a good opportunity to this end. After a very early start, we met at Elstree airfield to pick up the Aztec. The flight was a good experience for me also. Gordon let me take the controls for part of the flight and I soon found that my limited private flying experience was not quite up to the task of maintaining steady level flight accurately for long periods. Gordon was not too impressed and after a while he let the autopilot show me how it should be done!

We arrived mid-morning in the expectation of a quick briefing and a take-off early enough to allow us to make the return flight to Elstree that evening. We were greeted with the news that there was some hold-up with the transfer of the

TransMeridian Air Cargo McDonnell Douglas DC8-54 at Le Bourget, July 1977. (*Author*)

money for the purchase. Kevin Keegan (not the footballer), the chief pilot and a director of TransMeridian, was confident that it would be solved within hours. We went through our briefing and waited a while, but it soon became clear that we would not be flying that day, so we checked into the hotel where Kevin was staying. Over a few drinks and a good meal, I learned a great deal about the ins and outs of cargo airline operation. Kevin, the son of Mike Keegan (a big player in the cargo airline business), was starting out independently with his establishment of TransMeridian.

The comparison with passenger airlines was fascinating. The crew of the larger passenger airlines, such as British Airways, would arrive for their flight without having to worry about the management of the passengers. The performance calculations, the flight plan (route information), the amount of fuel required, the number of passengers and quantity of freight would all be determined beforehand at the airline's central offices and the information provided to the pilots. The freight and passenger baggage would be loaded by the ground crew in accordance with the central office loading calculations. The captain, being ultimately responsible, would check the information, familiarize himself with the flight planned route and prepare for take-off. In the cargo airline case, the Commercial Department would generally be responsible for obtaining cargo contracts. However, in contrast, it would be normal for the cargo airline crew to directly manage the loading of cargo. In common with the smaller passenger airlines, the cargo airline flight crew would often be responsible for formulating their own flight plan to the destination and calculating the take-off and landing distances, as well as rates of climb to ensure that runway distances and ability to clear any high terrain en route were acceptable. In addition to the flight-deck crew, out of necessity they would have to carry a loadmaster on the flight. In some cases, the crew would even be expected to find cargo loads for the return flight. Kevin summarized it in simple terms. When the cargo pilot had completed the take-off, the most stressful part of the operation was over, whereas the passenger airline pilot's work had only just begun.

Late the next morning the finance transfer was completed, and we were able to carry out the flight test. Everything was as expected and, after a short debrief, the UK Certificate of Airworthiness was duly issued in the afternoon and we were on our way back to Elstree.

Soon after this, another McDonnell Douglas aircraft type required CAA flight testing in Keith's absence: the Douglas DC-10-30. This was an interesting situation. The 10-30 had been certificated some years earlier, but it was found to be critically expensive for airlines to keep spare engines at their destination bases in case of engine failures en route. The reliability of these modern engines was significantly better than the earlier engine types, but they could still occasionally fail at inconvenient locations. When this happened, it was normal for a spare engine to be flown out, as cargo, to the airport where the aeroplane was stranded. However, it was not possible to carry these large fan engines in a normal cargo aeroplane. The solution was to carry one in a purpose-built pod, fixed under the wing of the passenger aeroplane in between the fuselage and the existing wing engine, often on the airline's next scheduled flight to the destination. Clearly, this had significant consequences for the aerodynamic and weight asymmetry, as well as the effect of extra drag on the take-off and climb performance capability. The inevitable reduction in available payload could be managed by limiting the number of passengers and/or freight carried, but if the aircraft was to be used for commercial operation it still had to comply with the airworthiness requirements. Apart from the structural implications of carrying the extra weight under the wing, the main potential problems were the flight handling/controllability requirements and, critically, the case of an engine failure on the same wing as the spare engine was carried. Having the drag of two engines on one side, the lowest speed at which the aircraft could be controlled was going to be higher than the normal situation

McDonnell Douglas DC10-30 with spare engine pod at Long Beach, August 1977. (*Author*)

without having the spare engine under the wing. As we have seen before, this means that the take-off speeds and take-off distances will be greater. This and other critical flight control issues meant a flight assessment by the CAA was essential; a trip to the McDonnell Douglas factory at Long Beach in California was arranged.

This was my first trip to the USA; another new adventure in the space of a few weeks. Gordon had other business in the USA before we were scheduled for the DC-10 flight testing. It was arranged that I would fly out to Los Angeles a few days after him. We would meet up with the McDonnell Douglas team at Long Beach to go over the company flight test results and plan our test programme. The British Airways flight to Los Angeles departed from Heathrow in mid-morning. The first surprise, for me, was to find that we were heading north for Scotland; I consulted the maps in the BA in-flight magazine and realized that the shortest ('great-circle') route took us over Greenland and northern Canada. Although it was a twelve-hour flight, because of the eight-hour time difference, it arrived in the early afternoon; plenty of time to get from the airport to Long Beach some 20 miles down the coast and start our work. Simple. However, there were two complications.

Firstly, Gordon's teenage daughter Bryony was taking the same flight to join her father for a short holiday with friends of Gordon in California. Gordon had asked me to 'look after her' on the journey. As I was in first-class and she was travelling economy, this was not straightforward. Secondly, due to some non-technical problem within BA, the flight had to be terminated in Chicago. All the passengers were shepherded into a transit area while BA sorted out bookings on connecting flights to Los Angeles for the passengers. Although there were several potential flights with different airlines in the next couple of hours, the re-booking process was not progressing very speedily. While as a first-class passenger I was expecting some sort of priority, I was concerned that Bryony might end up on a different flight. In reality, I think she was much more worldly than me about coping with these situations, but I decided it was time to go to the BA desk and explain the situation. The result was amazing. Within minutes, BA issued us both with tickets on the next Delta Airlines flight and shepherded us rapidly to the correct terminal with a promise that our luggage would also be redirected, and so it was. We arrived at Los Angeles airport later that afternoon. McDonnell Douglas had arranged to meet us from the flight and drive us down to Long Beach. However, although they were aware of the flight problem, they had no way of finding out on which flight we would be arriving. In my naivety, I assumed they would know, so we hung around in the arrivals area for a while before I again decided that I needed to take action. I had a McDonnell Douglas Flight Department telephone number that Gordon had given me, but I had no idea whether or not anyone at the other end would be able to help. To my relief and amazement, I was speaking directly to Phil Blum, the company test pilot who was with Gordon; a pick-up was rapidly arranged to everyone's relief.

On my arrival, although now very late in the afternoon, Gordon was keen to go through the test results and plan our flights for the next days. This was my first experience of jet-lag. It was one o'clock in the morning UK time; I had had a less than relaxed flight experience and was in a totally new environment. I remember having great difficulty concentrating on the flight test results that we were reviewing and the issues being discussed. Again, Gordon clearly understood my problem and gently shepherded me through. Later, back at our hotel during dinner, Gordon ran through things again. I never got the feeling that he doubted my competence in any way, but he wanted to be sure that I would be able to play my part to his satisfaction.

On the subject of jet-lag, Gordon passed on his advice: if you want your system to get over jet-lag as quickly as possible, stay up the first evening for as long as possible. His theory was that you would be so tired you would sleep for quite a long time, even though your natural clock would be wanting to wake you up before midnight local time. I tried this then and on later occasions, but was not convinced. No matter how tired I was, on the first and second mornings my 'time to wake up' clock was dominant and the later I went to bed, the less sleep I got. I adopted my own policy of going to bed at a reasonably early time on the first night, enabling as much sleep as possible before the inevitable early wake-up. After a few days, my clock would adjust, and normal time scheduling was possible. Subsequently, I found that the jet-lag problem was more difficult following a flight in a west-to-east direction. My time clock was noticeably more ready to extend the day than reduce it and after an eight-hour time change when the clock was put back, it could take me a week to get onto the destination time zone. This is quite a normal human situation, but some people's natural clocks are more biased in one direction than others. I remember one occasion, much later, when my second wife Elizabeth and I were travelling from Hawaii to London with a short transit stop in Los Angeles. From our different wake/sleep patterns over the next few days back in the UK, it was clear that Elizabeth's clock was struggling to go backwards while mine was going forwards, albeit needing to adjust by many more hours!

Back in Long Beach, I had no problem getting up the next day for an early flight. We took off at 08:15 for the first flight of three hours at forward c.g., followed, in the afternoon, by a second flight of two hours and thirty minutes at

Gordon Corps and MDC test pilot Phil Blum in DC10 with spare engine pod, August 1977. (*Author*)

aft c.g. As expected, the achievement of the engine-out minimum control speeds were the most difficult test points to confirm. The weight of the spare engine was compensated for by a clearly defined asymmetric fuel loading and usage schedule which had to be assessed. After a debrief and presentation of the CAA conclusions the following morning, I left Gordon for his short holiday and was back to Los Angeles airport for an uneventful BA return flight to London (arriving early the following morning, UK time).

Less than two weeks later, Nick Warner and I took the BA flight to Los Angeles to commence our three-week Lockheed L–1011 TriStar course. The Lockheed final assembly plant and Flight Test Center were situated at Palmdale in the Mojave Desert about 70 miles north-east of Los Angeles. As the BA flight arrived at about 2.00 pm, it had been the CAA Flight Department 'established convention' on previous visits to spend a couple of hours in the Wild Goose: a local 'club' near the airport, to 'unwind' before embarking on the drive to Palmdale. So Nick and I collected our hire car and, not wishing to break with tradition, duly found our way there. Our 'first impression' was unexpected. As we went inside from the bright Californian early afternoon sunlight, we were confronted with what appeared to be total darkness. It took quite a time for our eyes to adjust to the rather dim lighting before we felt confident that we could make our way to the bar! The Wild Goose could loosely be defined as a strip club, but most of the clientele were more interested in watching the large screens on the walls that continuously showed American football and baseball games as well as music videos of the current popular artists such as Meat Loaf. What all these people were doing here in mid-afternoon on a weekday was a mystery. By UK standards the beer was cheap and furthermore, every half-hour or so, free pizza was announced. So, after a couple of hours, feeling suitably refreshed, we set off to drive the 75 miles or so to Lancaster, the nearest 'city' to Palmdale. Nick did the driving and I was tasked with the navigation. Our map had little detail and not being familiar with some of the US signpost conventions, we made a couple of early mistakes. However, after less than two hours, we had safely arrived in Lancaster. As we were to be there for an extended period, Lockheed had arranged an apartment for us rather than the local motel. We had all agreed this would be homelier and it was also much cheaper. The apartment complex had a restaurant and breakfast was included in the price. Although Lancaster was quite a small 'city', we had some difficulty in interpreting the sketchy directions we had. However, we eventually located the apartment in time to unpack, have a quick dinner and an early night.

The next morning we easily navigated the short distance to the plant. At the security gate office, our passes were ready for collection. The Flight Center Chief,

Sam Wyrick, personally came to meet us and showed us the offices and lecture rooms where we would be spending the next couple of weeks. We soon got to know Sam very well. He was an impressive character; always very friendly and helpful, but he ran the Flight Center on a ruthlessly tight schedule. Planned programme overruns and wasted flying hours were not to be tolerated unless there was absolutely no alternative. Every problem and difficulty that arose – and, as is usual in any development and certification flight test programmes, there were many – was instantly brought to Sam's attention. He, with the support of his lead engineers, would burn much 'midnight oil', thinking inside and outside the box, to solve such problems to keep the programme on track. Even though he could be quite ruthless at times, we never met any of his staff who had a bad word to say about him.

We were introduced to the key test pilots – Bill Weaver, John Wells and Will Smith – and the lead engineers, as well as others who would be giving the training sessions. We were also given a detailed programme of the training. At the end, a few days had been set aside for Nick to directly familiarize himself with the flying characteristics of the TriStar. This would include some specific assessments of the more critical stability and control aspects. The aircraft to be flown was the L–1011–200 version and Sam was very keen to make sure Nick was specifically 'calibrated' on the certification standards of this type, which had been previously accepted by the CAA, before being asked to assess the later 500 version in a year or so.

We had expected to join others on the course, but it was soon clear that it had been specifically arranged for just the two of us. This was rather humbling. However, for most of these training sessions, visiting airline personnel and Lockheed staff joined us to take the opportunity to refresh their understanding of the specific subjects. All the people we met on this course – the flight test crews, the managers and engineers – knew John Carrodus and Don Burns very well and had nothing but praise for their work. We were made very welcome. In detail, we were taken through every aspect of the aircraft systems: the engine and fuel control systems, the hydraulic, electrical and mechanical flying control systems, the electrical generation and distribution systems, the pressurization and air-conditioning systems, the landing gear systems, the anti–icing systems and all aspects of the avionics and automatic flight control systems. I don't think I have ever been such an enthusiastically interested student. Looking back, in a relatively short period, that course taught me so much about the design philosophies and interactions of complex aircraft systems and, in particular, the way their failures could impinge on the flying qualities. This knowledge would be invaluable to me in many later certification projects.

At the end of the course, we took the examination; Nick because he would need it to get the TriStar on his licence (often a useful asset for a CAA test pilot to enable him to be an official flight crew member on commercial flights for route-proving assessments, etc.) and me, just for the satisfaction.

Finally, there was Nick's type conversion flying and the follow-up certification familiarization flights. The conversion flight went very well. To save time and money, Sam had planned to combine this flight with the final Lockheed flight evaluation of the new Flight Management System (FMS) before the FAA were to make their certification assessment in the next few days. At that time, this system was state-of-the-art. It had the ability to manage the en route climb profiles to minimize fuel use and it could also determine the optimum point at the end of the cruise at which to close the throttles for a final continuous descent into the airport terminal traffic pattern. We took off late in the afternoon so that night-flying could be included for Nick's experience. With the Lockheed pilots swapping places with Nick as necessary, everything was smoothly accomplished within a four-hour flight. With my new knowledge, I was entrusted to act as flight engineer, albeit with an experienced engineer keeping his eye on me.

The following morning, we took off for our certification familiarization flying but some unserviceabilities and very poor weather conditions caused a termination after forty minutes. Another attempt that afternoon also had to be curtailed. Now Sam had a problem. He needed to get the aircraft fully serviceable in a hurry and reconfigured for the important FMS flight test assessment by the FAA. Nick and I were now a bit of an embarrassment, but there was no way he wanted us to go home without the certification familiarization. From information passed on to us by John Carrodus, we knew well what Sam was worried about. In particular, the stall handling had been a difficult issue for John. There was not much natural nose-down self-recovering tendency at the stall and if not recognized, the pilot may continue to pull further back on the control. There would be no further reduction in airspeed, but a sudden wing drop could occur that could often exceed the 30 degrees bank limit for certification. On balance, and with the benefit of a second opinion from Dave Davies, it was deemed to be acceptable, but Sam was keen to ensure that Nick would concur and not raise it as an issue on the later models.

Sam got to work on a plan to give Nick and I something useful to occupy us for a few days. He used his influence with the Flight Operations management of Delta Airlines to arrange for Nick to act as co-pilot on an L-1011 TriStar commercial flight from Los Angeles to Miami and back. The normal co-pilot was also on board to be operationally legal. I would also be able to take the spare seat on the flight-deck. This would give us both a first-hand insight to a normal commercial operation of the TriStar. Sam did not want us back before they were ready, so we had two nights in a hotel with the dubious opportunity of experiencing Miami at the height of the high-humidity hot season before joining another crew on the return flight.

Back at Palmdale, Sam had everything fully under control again and scheduled a late-afternoon take-off for our flight. It turned out that there were a lot more items on Sam's programme than we expected, but it was all good experience for Nick and we covered all the areas with care. Nick's ability to fly accurate test points was second to none and this skill helped him to appreciate the validity of the final decisions made by John and Don in their assessments.

So that was the end of a very successful visit. We had made a lot of new friends and formed close working relationships that would prove to be invaluable for our progress through some of the difficult issues with which we were going to be faced in the future.

Chapter 11

Never a Dull Moment

After I returned from the TriStar course at Palmdale, on the programme for me were a couple of flights with Dave Davies and a day trip to Dublin with Gordon Corps for an airworthiness flight test on a Learjet. The first of the two with Dave was an old Series 1 748 imported by Dan-Air from Aerolíneas Argentinas. It needed to have a series air test as part of the inspection and acceptance process before it could gain a UK Certificate of Airworthiness. Dave was less than one year from retirement and, throughout his long career as the CAA's chief test pilot, I think you could count, on the fingers of one hand, the number of times he had flown a 748. Nothing fazed him; you would think he had flown it every day. In some ways, the nature of our airworthiness flight tests was so different from normal airline flying or training that the Dan-Air captain in the co-pilot's seat would have been quite impressed with Dave's capability even if his execution of the stability, control and stalling checks had been less precise. I, however, was qualified to be impressed. This 748 was showing its age; the climb performance was marginal as were the usual problem handling areas, but it did pass and a Certificate of Airworthiness was duly issued by the local CAA surveyor.

At the other extreme, the second flight with Dave was an airworthiness flight test on an Air Malawi BAC (originally Vickers) VC10. British Airways had a contract with Air Malawi to carry out all their routine maintenance and this included the C of A renewal air test. Dave was a big fan of the VC10 and found it particularly satisfying to fly. There were no problems, and the CAA was able to recommend to the Malawian Aviation Authority that they could re-issue their C of A.

Back from his Lockheed TriStar L-1011 flying 'training', Nick was keen to maintain his familiarity with the type. Whenever the pressure of other work allowed, he took several opportunities to fly the airworthiness flight tests on the British Airways L-1011s. During these flights, we got to know the BA crew very well. Terry Lakin was the BA TriStar fleet captain, his right-hand man on the air tests was a senior first officer Dave Cretney, and their regular flight test engineer was John Webster, who I had met regularly on earlier Trident airworthiness flight tests.

On one such occasion, it was a Gulf Air L-1011 we were flying. Similar to the BA contract with Air Malawi on their VC10s, BA looked after the major maintenance of the Gulf Air L-1011s. During this test flight, we found a major problem. A 'standard' test on the CAA Airworthiness Flight Test Schedule (AFTS) was a check of the ability to re-start an engine in flight. It was an airworthiness requirement to be able to re-start an engine within a defined envelope of speed and altitude conditions (the re-start envelope). In this case, it proved impossible to re-start one of the engines except at the higher speed end of the re-start envelope. After the flight, the BA ground engineers made some adjustments and the next day, we tried again. There was an improvement, but it was still not possible to re-start the engine at the lower speeds within the certificated envelope. BA had to enlist the help of Rolls-Royce to locate and fix the problem. We left it to BA to carry out the subsequent re-test and it re-started OK. It was a good example of the justification for these airworthiness flight tests; identifying a problem that would not otherwise have manifested itself until an emergency situation arose on a normal in-service flight.

During this period, the CAA needed to carry out a series flight test on the first Rockwell 690A imported to the UK. There were no CAA test pilots available at the time, so we agreed that Dave Gollings would do the flying with me as flight test engineer. Dave was a key part of the UK Rockwell importing agency, Glos Air, based at Bournemouth Hurn airport, and an experienced pilot on the type. This should have been a straightforward day out for me. I arrived in the morning, went through the test programme with Dave and we took off at about midday. Immediately after getting airborne it was clear that something was seriously wrong with the altitude and airspeed indications. Dave made a cautious circuit of the airfield and landed back on the runway using his experience and giving some credit to the indications of the standby airspeed indicator without being totally sure how accurate it was.

A leak in the primary static pressure line to the airspeed and altitude indicators was soon diagnosed and fixed. After taking the opportunity for some lunch, we took off again at about 15:00. The first item on the programme was the engine-out climb performance check. After the first five-minute climb, my calculations showed that the rate of climb was significantly below the scheduled performance in the Flight Manual. We repeated the test a further two times with only a small variation in the result. There was little point in continuing with the remaining flight tests with this problem existing, so we returned to Hurn. A good look around the aeroplane suggested that the wing-flaps were slightly down from their fully retracted position; this would create some additional aerodynamic drag and

Rockwell 690A, G-BEJN at Hurn, 1977. (*Author*)

could possibly account for the problem. The engineers made some adjustments to the flap linkage and at about 17:00 we were ready to try again. This time there was some detectable improvement, but the performance was still below an acceptable tolerance. Time to call it a day and leave the Rockwell agency to look further into the problem; a long day by the time I arrived back home.

I discussed the situation back at the CAA with Keith Perrin, Andrew McClymont and the test pilots. I was able to assure them that I had no doubts about the accuracy of Dave Gollings' flying and we agreed to continue as before. Four days later, I was back at Hurn to try again with Dave Gollings. Glos Air had consulted Rockwell in the USA who had no suggestions to offer other than checking the engines and looking further for anything that would increase the drag, such as protruding access panels. Some slight adjustment to the engines, as well as minor improvements to the fitting of landing-gear doors and some access panels, had been made. We took off to try again. The engine-out climb was the first thing to check. We did two five-minute climbs in opposite directions to even out any effects of atmospheric variability, but the results were little different from the previous flights. We continued with the remainder of the test programme without any further problem. After the flight, we had a debrief. There could be no argument that the Rockwell scheduled climb performance appeared to be optimistic and the CAA would require a performance correction to be included in the UK Flight Manual. In fairness, I suggested that we should make another flight early that evening, when the atmospheric conditions were more stable. We carried out a further four climbs to get as good a set of data as possible. The average result was as good as we had seen and consistent enough to enable Andrew McClymont to review the performance with Rockwell. A rate-of-climb correction was subsequently added to the UK Flight Manual. Uncovering this problem was another good example of the value of these airworthiness flight tests.

By now hand-held calculators were becoming commonplace and on my previous visit to the USA, I had found just what I needed to help manage the in-flight calculations. It was a Casio 'scientific' calculator only slightly bigger than a credit card. It did everything I needed including holding my most-used conversion factors such as pounds to kilogrammes, miles per hour to knots, and litres to gallons. I fashioned a suitable fitting in which to slide the calculator and it sat conveniently on the opposite top corner of my clipboard to the stop-watch. It worked perfectly for the rest of my flight-testing career. Sadly, it became irreparably damaged sometime later; I was never able to find anything remotely similar in size and capability.

At about that time, we were joined by a new flight test engineer, John Denning. He replaced Don Burns. John had gained his flight test experience at Boscombe Down on military aircraft, but his enthusiasm and desire to learn about civil aeroplanes was second to none. After a few months in post, he was allocated responsibility for the Airbus Industrie aeroplanes; this included their new 'fly-by-wire' A300. Although not an unusual concept for military fast jets, it was new to the civil aircraft world. John had a very sharp mind which, coupled with his experience, made for a formidable team with Gordon Corps, who was the CAA project pilot on the Airbus types. Over the following years, through Airbus developments of several new types, Gordon certainly appreciated John's contributions to the difficult certification issues that arose from this concept. However, probably because they were both similarly forthright characters, their team-working was a little strained at times.

With the addition of John Denning, we were back to a complement of four test pilots – Dave Davies, Gordon Corps, Nick Warner and Darrol Stinton – and four handling flight test engineers – Keith Perrin, Dave Cummings, myself and John Denning – as well as the performance flight test engineers – Andrew McClymont and Bill Horsley – which was an appropriate number for the increasing workload of the fixed-wing team. On the helicopter side, the flight test engineer Mike Smith had been working with Ken Reed. Ken was a fairly elderly test pilot, working for the CAA on a part-time contract. The number and variety of helicopters was increasing rapidly, such that the CAA Flight Department decided they needed a permanent test pilot. In 1979, Peter Harper, an ex-naval helicopter pilot who had graduated from the test pilots' school at Boscombe Down, was recruited.

In December 1977, Nick Warner and I, with Peter Richards and Denis Murrin were back in São José dos Campos to evaluate the Embraer EMB 110K1. This version had a large cargo door in the rear port side of the fuselage instead of the normal passenger entry door. Additionally, there were several other changes to the previous versions we had assessed, some of which were specific CAA required modifications. Most of these Embraer considered to be design improvements and they had introduced them, as production changes, to all civil versions. The aeroplane's systems were unchanged so the only aspects to be assessed were the flying qualities and the structural implications of the cargo door. Denis coordinated the whole process. Peter Richards was our structural specialist for the type and was, in common with many of our engineers, a real 'character'.

As before, we had arrived by Concorde with Embraer's full agreement, in spite of their continued insistence of a night stop in Rio thereby negating any saving of time in the programme that would justify the extra air-fare cost to them. It was Nick's first visit to Embraer and the four of us gelled extremely well, both technically and socially.

During this visit, Embraer had also invited senior representatives from their UK importing agents, CSE Aviation at Kidlington, Oxford. The CSE managing director at the time was Lord Waterpark. In discussions with Denis Murrin before the arrival of the CSE personnel, Embraer were referring to him as Mr Lord; it dawned on Denis that they had not appreciated he was a 'real' lord. When Denis tactfully appraised them of the situation, the Embraer Sales Department corporate hospitality machine leaped into overdrive. Inevitably, Denis was caught up in this from time to time which, although pleasurable, put pressure on Denis's programme timing. Nick and I kept our heads down in the Flight Center and were largely left to get on with the flying.

As an interesting story, several years later, after I had left the Flight Department, I continued to be closely involved with Embraer on the UK certification of their subsequent projects. At the Farnborough Air Show, they had invited me for lunch in the Embraer chalet for discussions. My wife Elizabeth, who also worked for the CAA, was with me at the show and they had included her in the invitation. We had just sat down for lunch with Embraer and a couple of other CAA people when Lord Waterpark, who was also in the chalet, spotted me and came up to the

Denis Murrin (head of CAA Foreign Aircraft Certification) with Embraer chief engineer (on left), São José dos Campos, December 1977. (*Author*)

table. His first name was Carol; an unusual but not unknown name for a man. His friends and close colleagues addressed him as 'Cal'. He clearly wanted to talk to me, so I moved my chair back ready to stand up. Swiftly, he said he did not want to disturb our lunch, but he was about to go into a sales meeting and needed to know about the resolution of our certification issues. He squatted down between my chair and my wife's, introduced himself to my wife and continued with his question to me. I was able to assure him that the additional testing and analysis Embraer had done had closed the issues. He went away happily to his meeting. On the way home, my wife remarked that it had been a surreal experience to have a lord kneeling down in front of her!

The two most important areas to be assessed during our flight tests were an extension to the forward c.g. limit and the autopilot. Because of the lengthening of the fuselage nose, Embraer found it was all too easy to load the aeroplane forward of the existing forward c.g. limit, especially if most of the passengers were seated towards the front. Although the elevator control forces were a little greater at this new forward limit, Embraer had satisfied themselves that the worst cases were still

Embraer EMB 110K1 with Nick Warner, Cabral and the author, São José dos Campos, December 1977. (*Author*)

just within the maximum limits of the requirements. The autopilot was a later version of the one installed on our previous flights.

With Cabral, the Embraer test pilot, again in the right-hand seat and Nick doing the flying, we went through our normal gambit of critical flight tests and found that all was well. We concurred that the critical cases of elevator force limits were just achievable at the new forward c.g. limit. Then we had to assess the autopilot. During our earlier flights, when en route to and from the test areas and in between tests, Nick had made use of the autopilot. It controlled the aeroplane very well, so our remaining testing would be limited to assessing the failure cases. With the aeroplane loaded to the aft c.g. limit, critical for the failure cases, we set off towards the coast. The Embraer airfield was in a mountainous area. Generally, we preferred to do most of the testing over the sea where the air was less disturbed and, if necessary, we could operate at lower levels where, in the denser air, the engines were generating their maximum power. It was less than twenty minutes' flying time to the nearest coast. On our way there, Nick noticed that, after he had engaged the autopilot with the ailerons carefully trimmed for wings-level flight, each time he subsequently disengaged it, the aeroplane would roll noticeably to the right. This was not normal and my nervousness about critical autopilot failure testing cut in. I suggested we should not be attempting such testing with this unknown problem. I detected a little hesitation in Nick and Cabral, but there was no disagreement and we turned back.

The autopilot was thoroughly tested and showed no faults, but an examination of the wing trailing-edge revealed that one of the fixed 'trimming' tabs was bent noticeably more than normal. This was straightened and off we went again later the same day. The problem was no longer evident, and we went through all the hard-over and oscillatory failures without finding any problems. I still found it difficult to understand why, once trimmed for wings-level flight, the autopilot would somehow ignore the extra manually-applied trim and compensate with its own applied correction which, when disconnected, resulted in the aeroplane rolling.

Sadly, we thought that this might be our last visit to Embraer for some considerable time, but they had another project in mind for UK certification and, less than a year later, we would be back.

Back to reality in Redhill, it was report-writing and several months of airworthiness flight tests, mostly Tridents and L-1011 TriStars, but including a couple of Dan-Air 748s with their 748 fleet chief pilot, another John Smith!

The CAA had received several Type Certification applications from foreign manufacturers, and Denis Murrin was hard at work planning a series of foreign aircraft validation visits for the CAA teams. The next on his agenda was the Piper PA-31T1 Cheyenne. It was an upgraded version of the piston-engine Piper Navajo in which I had previously flown a few times on airworthiness flight tests with Gordon Corps. However, it was now fitted with Pratt & Whitney PT6A-28, 620 shp turbo-propeller engines. Being turbine-powered, it now came into the category of foreign aeroplane types for which the CAA needed to carry out a design investigation.

At that time, the Piper main factory was on an out-of-the-way airfield situated at Lockhaven in Pennsylvania. There were no scheduled passenger flights into Lockhaven, so Piper had arranged to meet us off the British Airways flight into New York JFK and ferry us to Lockhaven in two of their aeroplanes. Gordon Corps was already in the USA, having been with Keith evaluating the latest Grumman Gulfstream 2 business jet at Bethpage in New York State and would make his own way from there by car. The remaining seven of our team would take the Piper flights. After clearing immigration and customs at JFK we were shepherded with our bags out through a side door in the terminal building, straight onto the tarmac where their two aeroplanes were waiting; one a PA-31 Navajo and the other a smaller PA-23 Aztec. For some reason, their two pilots were only expecting six of us and it was clear that they were also surprised by the number and weight of our bags. For visits of ten days or so, and with all the paperwork each of us needed, one large heavy suitcase and one big briefcase or flight-bag was our normal baggage. The flight to Lockhaven was about one and a half hours and they needed plenty of fuel. Those of us familiar with such situations realized that the aeroplanes were going to be close to their maximum weight and with all the baggage in the rear compartment, probably at or just outside the aft c.g. limit. These observations were substantiated when the pilot of our Navajo made us sit in the forwardmost seats and even strapped some of the suitcases onto passenger seats.

The take-off run was a bit laboured, but the length of the JFK runway was more than adequate. With both engines at maximum continuous power, the rate of climb was not too bad, but we soon realized that we were heading for a somewhat mountainous region and we certainly needed to be climbing. It was not long before we entered cloud, still climbing, and the pilot was carefully checking our position on his map using the transmissions from the nearest available navigation beacons (many years before the magic of GPS). At some point, he was going to have to descend through the cloud to reach the airfield at Lockhaven. From time to time, he was in communication with the Aztec behind us to confirm our relative positions. Both pilots were obviously familiar with the terrain and had been in this situation before, so we probably should not have been worrying quite so much.

Piper Airfield, Lockhaven, Pennsylvania, May 1978. (*Author*)

When the time came to descend through the cloud, there below and in front of us was the airfield. It was quite turbulent and the twitchiness of the aeroplane at aft c.g. gave the pilot a good test of his skill on the landing.

Piper had a couple of hire cars ready for us and we made our way to our hotel. It was not a very salubrious establishment and the rooms were quite scruffy and pokey; what a contrast to the standards I had experienced in Brazil and California. It was late afternoon and Gordon Corps had not yet arrived. We all crowded into Denis's room for a discussion, and it was clear that I was not alone in my assessment. Those in the team who had been on more foreign evaluation visits than I were not at all happy; apart from the general scruffiness, there was nowhere in the rooms to write our individual reports. Denis made one of his instant decisions and found that there was a Holiday Inn only a little further away from the airfield, but he wanted to wait for Gordon to arrive before setting any change in motion. Although Denis had a lot of authority and flexibility in making choices such as suitable hotels, in this case, as it was Piper who had made the hotel booking, Denis, before taking any action, was keen to have the full support of Gordon, being the next most senior person on the team. Gordon arrived soon after and it took him seconds to assess the situation and give Denis the support he needed. We decamped to the Holiday Inn that evening and the team morale went up by several notches.

The difficulty was that, next morning at the initial introductory meeting, Denis had to explain our actions to the Piper people in charge of our visit. Things were a little frosty at first, but some of the Piper personnel who had experienced previous CAA visits understood and this alleviated our embarrassment. As was customary, we began with the introductions of our team members and the Piper specialist engineers assigned to be working with us. Apart from Gordon, myself and Denis, we had with us Andrew McClymont (Flight Performance and Flight Manual), Mike Benoy (Structures), Bert Rust (Electrical and Avionic Systems) and Vin Wills (Engine Installation and Fuel Systems). As usual, with the simpler aeroplane types, Denis would cover the Flying Control Systems and Cabin Safety issues. (On this particular visit, we also had with us a trainee understudy to Vin, whose presence was entirely funded at the CAA's expense.) We then had a general overall presentation of the design features of the Cheyenne to give us all a good overview and an insight into aspects where more than one engineering discipline may need to interact. The problem for the flight-test team was that in some cases we wanted to know more detail about the system being described in this 'summary' presentation in order to determine whether failure cases needed to be evaluated and included in our flying programme. Gordon usually wanted all the answers straight away, which meant he would ask a lot of detailed questions. This tended to irritate some of our specialists who thought he was trying to do their jobs for them, and it was significantly prolonging this initial meeting. I thought we had sufficient information to be going on with and any necessary supplementary information could easily be gained later. Denis was used to Gordon and I was watching him to see how he would wrest back control of the meeting. Denis looked across at me and I raised my 'slightly impatient' eyebrows. It was all the cue Denis needed; he looked at his watch with a flourish and demanded that we move on to keep to schedule, asserting that such detail could be dealt with later.

Once we got into face-to-face discussions with our opposite numbers, it was clear that the Piper mindset was much less accommodating than Embraer, Douglas or Lockheed. They did not want to know about problems, they were looking for ticks in boxes. We went through their flight test reports and results to confirm where the critical stability and control cases were. In common with most aeroplanes of this class, where piston engines had been replaced with more powerful turbine engines, everywhere we looked, the results appeared marginal. The longitudinal stability had been enhanced with the addition of a 'down-spring'. As explained earlier, this was a spring incorporated into the elevator control circuit, which acted to increase the control force so that when the control was released, this residual force helped to bring the aeroplane back to the equilibrium state. A bob-weight was also fitted into the elevator control circuit to increase the stick force per g. There were also two variants they wanted us to certificate: with and without the installation of wing-tip fuel tanks. Nevertheless, they were a bit surprised about the amount of testing included in our flight programme; we included functioning checks of the emergency landing-gear lowering system and the emergency pressurization. They also seemed a little hesitant to let me have a copy of the weight and balance data which I needed to check the weight and c.g. position in flight. Sometimes, when it was critical for a particular test to be exactly on the aft c.g. limit, I could work out whether I needed to move temporarily to a more aft seat or location in the cabin.

The Piper project test pilot for the Cheyenne who flew with us was a likeable, if slightly dour, chap called Bill Lawton. He never tried to unduly influence us; he answered all our questions in a straightforward manner and on two or three occasions, uttered the words: 'I have never seen that before!'

Such a response had often been said by other company pilots on similar occasions when we encountered a problem and it elicited a brief eye-contact between Gordon and me.

Piper PA–31T1, Lockhaven, May 1978. (*Author*)

Gordon Corps and Piper test pilot Bill Lawton in Piper PA–31T1, May 1978. (*Author*)

We completed our programme in about nine hours with six flights; this included both tip–tank configurations, as well as a night flight to assess the cockpit lighting and warnings in the dark. We finished the flying late on the Friday afternoon, so we had the weekend to prepare our conclusions for our formal debrief on Monday morning. Clearly, it would not take us the whole weekend, so Gordon suggested we could drive to Gettysburg to see the site of the battle of the American Civil War. Although no more enthusiastic then about history than in my school days, I happily agreed as I was interested to see more of the country and a bit of culture should do me no harm.

It turned out to be an unforgettable experience. It was about a two-hour drive through picturesque countryside. On arrival, we first went to the Visitor Center. In contrast to any similar historic site in the UK at that time, it was a perfect lesson on how things should be done. There was a big three-dimensional display map of the whole battlefield area with all the important events identified. A very helpful and knowledgeable young man (probably a college student) walked our small party around the display and explained all the phases of the battle in some detail. We were then given a cassette recorder that we could plug into the car's cigar-lighter socket, and a map with the directions for a drive around the site with places to stop. At each stopping-point, we started the recorder and it explained everything that happened in that location. It was so fascinating and well done – in total contrast to school history lessons – that I instantly gained an interest in history that never left me. In the Visitor Center, I purchased an extensively-illustrated book covering the whole day-to-day history of the American Civil War. Instead of trying to get some sleep on the flight back home from JFK, I read it avidly.

Denis Murrin always tried to plan these team visits so that he had a weekend at the end where he could collate all the reports and conclusions of the individual team members and give his final overall presentation to the company's senior people on the Monday. Each team member was expected to write a description of the aeroplane's systems or structure in the areas of their responsibility. It had to be in sufficient detail to explain the reasoning for accepting, or not, the design features against the relevant BCAR requirements. Appended to this would be their lists of conclusions, divided into the following four categories:

1)   Unacceptable items.
2)   Areas where some additional technical information was required before a decision could be made.
3)   Areas where the letter of the requirement was not met but there were compensating features so that a finding of 'equivalent safety' could be made.
4)   Criticisms of features where the requirements were met but which the team believed could be improved upon to the betterment of the type.

Denis did not really like the subjective idea of 'criticisms', but it was a long-standing Flight Department belief that they could be of value when the manufacturer introduced an upgrade to the model or a new variant. This did sometimes happen. Occasionally, the other team members would add a criticism to their own conclusions. The flight team's categories included 'major criticisms' and 'criticisms'. Denis treated the lower category of 'criticisms' with some disdain and rarely mentioned them during his debrief.

Each team member would have his report and conclusions typed up by the manufacturer and Denis would go over each with his experienced eye for inconsistencies and any areas where he thought some additional inter-discipline consultation was necessary. Each member would debrief both their report and conclusions with their assigned company specialists to ensure their full understanding of the issues. This way, when Denis gave his overall team presentation at the end, there should be no surprises for the assembled company people, but invariably there would be some controversy over one or two of the CAA's interpretations.

Gordon and I had completed our conclusions on the Saturday and handed them to Denis before a rather boozy supper in the Holiday Inn restaurant. While Denis was finishing his collations the next day, we were taking in Gettysburg.

It was usual for the team members to have finished and departed before Denis's debrief. In this case, Gordon and I were still there, and Denis was able to sit in on our debrief first thing on the Monday morning. After a short break, he was able to give his. Gordon and I sat at the back of the room. It was a masterly demonstration of authority, tact and understanding of the manufacturer's position. He dealt with the awkward questions calmly and with confidence.

We left Piper with a few problems to ponder over, and the three of us drove the two to three hours back to New York to get the evening flight to the UK. On the way back, I learned from Gordon about a near-catastrophic incident he had while flying the Gulfstream 2 the previous week before the Piper evaluation. Gordon was flying from the left-hand seat as usual. During a normal take-off, they encountered a flock of large birds, some of which entered both engines just as the aeroplane was about to get airborne. At that stage in the take-off, they were above the decision speed (V_1) and strictly should continue the take-off. Gordon made an instant decision, as he had no idea whether either or both engines were still capable of producing significant thrust; he closed both throttles, put the aeroplane firmly back on the runway and stood on the brakes. Even at that speed, there was just sufficient runway left to complete the stop. I heard later that the Gulfstream test pilot and others on the flight-deck burst into an impromptu round of applause. Apart from two damaged engines and some very hot worn-out brake pads, they had Gordon to thank for still having themselves and their precious aeroplane intact.

Back in the UK, Gordon and I had a couple of flights on Trident aeroplanes at Hatfield. We had to check the aerodynamic effects of the wing-strengthening modifications that were necessary after a significant miscalculation had been found in the structural analysis. This was discovered several years after certification approval. It was, therefore, extremely embarrassing both to HSA and the CAA.

Next, Denis had organized another team validation visit to Embraer. When last there, Embraer had shown us their Xingu (pronounced 'shin-goo') prototype. It was a corporate-type aeroplane that used the wings and engines from the Bandeirante with a smaller fuselage for up to seven occupants. We had thought that, with so many similar types currently on the market (the PA-31T1 Cheyenne being one such example), there would be a limited market for it in the UK. However, Embraer believed there would be enough UK customer interest and

Embraer EMB 121 Xingu prototype, São José dos Campos, June 1978. (*Author*)

they particularly valued our opinion, so we were asked to make our validation assessment; in this case before the FAA.

As all the systems were almost identical to the Bandeirante, it was quite a select team that embarked on the now well-trodden path to Embraer at São José dos Campos. It comprised Denis, Nick and myself with Peter Richards to deal with the structural aspects, Bill Horsley to handle the scheduled operational performance and Bert Rust to cover electrical and avionics systems. Denis was perfectly capable of assessing any other minor systems differences from the Bandeirante.

In common with most of these short fuselage, twin-turbo-propeller corporate aeroplanes, the minimum control speeds and the stall handling were always difficult areas. In addition, as with the PA-31T1 Cheyenne, the Xingu needed a down-spring for longitudinal stability and a bob-weight to meet the stick force per g requirement. They had also found it necessary to fit a stick-pusher as the 'T' tail configuration had brought with it a 'deep-stall' tendency. All the critical

Embraer Xingu with Embraer flight test engineer Hyodo, Nick Warner, the author and Cabral, São José dos Campos, June 1978. (*Author*)

Embraer Xingu cockpit with additional test instruments fitted above coaming, June 1978.
(*Author*)

certification areas were showing up compliance problems. We did a lot of stall testing, both at forward c.g. and aft c.g. to check the operation of the stick-pusher as well as Embraer's declared stalling speeds.

We went away, leaving Embraer with quite a list of problem areas to ponder which included some cases of rudder over-balance. Although they were unsurprisingly disappointed, they were grateful to have some time to address these issues before inviting the FAA to conduct their validation. The USA was the largest potential market and achieving an FAA Type Certificate was Embraer's primary goal.

The last half of 1978 was filled with an assortment of airworthiness and series flight tests: several versions of the HS 125 business jet with various production modifications; a couple more Dan-Air 748s with the other John Smith; and a number of small aeroplanes with Darrol. These included a series check on a new Piper single-engine variant, the PA-32RT, but the most memorable was a brief but intense flight test of the new NDN-1 Firecracker in order to grant a Special Category Certificate of Airworthiness. This would enable the NDN Company to carry out demonstration flights outside the UK before their certification programme had been completed. The NDN-1 Firecracker was designed by Nigel Desmond Norman, one half of the Britten-Norman Aircraft Company

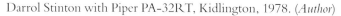

Darrol Stinton with Piper PA-32RT, Kidlington, 1978. (*Author*)

founders. He had set up NDN as a new company with the intention of offering the Firecracker as a cheap option for training military pilots, aimed at third-world countries. It was powered by a Lycoming 0-540 piston engine. More of this later.

In early 1979, back in tandem with Gordon Corps, we went to Istres in the south of France to carry out the validation flight testing of the Dassault Falcon 20F business jet. The 20F was a new development of the 20E, which had been previously certificated by the CAA. The major change was the addition of wing leading-edge slats to significantly reduce the take-off and landing distances. Apart from my pleasant surprise at the warmth of the weather in February, we encountered few unforeseen surprises from the content of our typical six-hour flight-test programme.

A month later we were invited back to Embraer to re-assess the Xingu after some modification work to solve our difficulties. They had increased the strength of the elevator down-spring, added a ventral fin to improve the directional handling, as well as vortex generators ahead of the ailerons and fin. Embraer now had a new (to us) test pilot on this project, known as 'Sergio'; very competent, but not as cheery

Dassault Falcon 20F with the author, Dassault test pilot Pierre Raisal and Gordon Corps, Istres, February 1979. The gentleman on the right was a Dassault engineer but unfortunately his name is unknown. (*Author*)

Embraer Xingu showing vortex generators ahead of ailerons, February 1979. (*Author*)

as Cabral. Initially, we flew the Xingu prototype as we had on the last visit.

The first flight was primarily to re-assess the stall handling and stall speeds with the stick-pusher. There was no problem with the pre-stall handling, but we were having difficulty achieving the Embraer scheduled stall (stick-pusher operating) speeds. There was some concern that the airspeed indicator might be over-reading at the lower speeds. Embraer obviously wished to investigate this, and so for our next flights we used a new production Xingu. It was partially equipped internally, so we were only able to do the testing at the higher weights. Everything went well, including the stalling speed checks, and we accomplished most of the

programme. However, after three flights,

Embraer Xingu showing vortex generators ahead of rudder, February 1979. (*Author*)

Pitot-static probe fitted to Embraer Xingu ahead of wing-tip for accurate measurement of airspeed and altitude, February 1979. (*Author*)

we were back in the prototype. Embraer had found a problem with the pitot-static system of the prototype and, although they believed they had corrected it, they elected to re-install the pitot-static test probe forward of the wing-tip in order to accurately check the normal airspeed indications. This probe is usually only installed, at the beginning of the development testing of a new prototype, for the purpose of accurately calibrating the normal flight instruments. With this, we completed all the low weight, aft c.g. critical tests without a problem. The Xingu was the second of two Embraer types on the way to UK Certification in the space of less than two years.

Another aspect of this visit, which we had not realized before our arrival, was the famous Rio Carnival; it would be taking place during our stay. Embraer had made arrangements for us to experience the atmosphere and watch the actual parade; a further inevitable but extremely entertaining disruption to our programme! Unlike the American FAA, who had strict rules about accepting hospitality and gifts from manufacturers and airlines, at this time the CAA was less circumspect. Getting to know the company's people on a more personal basis was a tremendous help when negotiating difficulties. I never saw, or heard of, any CAA person compromising a technical decision as a result. In fact, in some areas, we gave them a harder time than the FAA. It also appeared to have been a very sore point with Embraer that the FAA refused all the renowned Brazilian hospitality that they had been offered.

A little more than a month later, we were about to embark on what Nick and I had been most looking forward to: the Lockheed L-1011-500 TriStar evaluation in Palmdale, California. It transpired that the week before our visit, Gordon Corps and Keith Perrin were due to be at McDonnell Douglas in Long Beach to do a pre-delivery airworthiness flight test on a new Laker Airways DC-10. Keith thought it made more sense for me go out a few days early with Gordon rather than make the return trip himself. This we did. McDonnell Douglas had much earlier adopted the policy of having the UK Certificate of Airworthiness issued in the USA before the aeroplane's delivery to the UK. Experience had told them that the cost of having the necessary CAA people come to Long Beach where McDonnell Douglas had all their technical expertise to deal with last-minute problems was a cheaper option than trying to sort out problems in the UK.

We arrived at Los Angeles Airport (LAX) in the early afternoon and drove straight to Long Beach for the flight briefing. Due to the local commercial traffic and the crowded nature of the surrounding airspace, McDonnell Douglas had difficulty carrying out anything other than short test flights from Long Beach. Consequently, they had set up a flight test facility at Yuma, an airfield in Arizona, close to the Mexican border. This is where the DC-10 would fly to the next morning with us on board. An early start was necessary; no problem for me and Gordon with jet-lag. We arrived at Yuma about 10.00 in the morning (one hour back from the West Coast time zone). I remember standing in the open aeroplane doorway ready to walk down the steps, being conscious of a searing dry heat. For a brief moment, I thought some internal cabin heater was malfunctioning. I soon realized that this was the local ambient air temperature of Yuma (about 100 degrees F and very dry). It took some getting used to.

While the aeroplane was prepared for our flight, we had a small snack and took off at 13:45 local time. It turned out to be a long flight because McDonnell Douglas had quite a long list of their own checks to do and it was late by the time we landed, feeling quite tired. Back at the hotel, the McDonnell Douglas flight-test people were not ready to leave us on our own. After a session of 'hoovering' down several margaritas – which were delivered in pitchers – we all went into the restaurant for dinner. After that, sleep was immediate, but it was not very long before my internal time-clock kicked in.

The next day we had the debrief, flew back to Long Beach and returned to Los Angeles in Gordon's hire car. Gordon's flight to London was the following afternoon so we checked into a convenient hotel. The next morning was Tuesday. Nick was due to fly into Los Angeles on Thursday as Lockheed had planned, unusually, for us to do our first flight on a Saturday. They wanted to use the much larger runway at nearby Edwards Air Force Base for the Minimum Control Speed tests (V_{MCG}); because it was a military base, the weekend was the only time it would be free to operate from without interruptions.

Gordon had to return his rental car before his flight. After Gordon had left, I needed to have a car for me to use with Nick. Gordon and I drove together to the car rental company so that I could get my own car. Somewhat amazingly, a car of the size we had ordered for the Thursday was not available and at that time of day and mid-week they had very little choice. To my surprise and delight, they offered me a Chevrolet Camaro for the two days and I accepted it. Gordon was a little jealous. We drove in convoy to a place for lunch that Gordon knew and recommended. It was a delightful restaurant on the edge of the Marina del Rey called the 'Warehouse'. I have been back there several times since. After lunch, Gordon went off to the airport and, after lingering there a while longer, I returned to the hotel. I was ready to spend the rest of that day and Wednesday catching up on report-writing and preparing for the Lockheed visit. The weather was beautiful, and I had a re-think; why not take the opportunity to explore the mountainous scenery to the east of Los Angeles? I looked at the road map and saw that it was less than a two-hour drive to the foot-hills and decided that a day's leave was called for. Next day, after an early breakfast, I was in the Camaro snaking my way along convenient freeways and out of the sprawling city onto small roads. The spectacular mountain scenery opened up; I stopped several times to view and take photographs. At lunchtime, I came across a delightful café, a shack overlooking Crystal Lake. Eventually, I returned to the hotel feeling very pleased with my re-think.

The next morning there was time for some report-writing before returning the Camaro and exchanging it for something less sporty that would accommodate Nick and I with all our luggage. I met Nick from his flight, spent our customary couple of hours in the Wild Goose and drove towards Palmdale. Lancaster was the nearest sizeable 'city', where we checked into the Lancaster Motel.

The friendly welcome we received next day at Lockheed was just as we had remembered from our previous visit. There was the cheery, crew-cut Sam Wyrick ready to go through his plan for our visit. He had second-guessed almost everything in our programme for the 500 and was ready to show us their reports and analysis of the critical test points. It had to be, and was, a productive day for us all. The Saturday morning flight next day was scheduled for a 06:30 departure, and

Chevrolet Camaro hire car, Marina del Rey, Los Angeles, March 1979. (*Author*)

they hoped to get two flights into the day. We did the stalls and some of the aft c.g. stability checks before the aeroplane weight was low enough for the V_{MCG} testing at Edwards Air Force Base, then back to Palmdale for refuelling. Due to the shorter fuselage of the 500, the fin and rudder are closer to the centre of gravity, and consequently directional control with an engine failure is more critical. Minimum control speeds are invariably the losers in this situation and the determination of V_{MCG} is always one of the most demanding tests. The reader may recall, from the 748 situation in Chapter 5, that it requires rapid and precise responses to control the lateral deviation following the engine failure and the accurate achievement of scheduled rotation and lift-off speeds. We completed the second planned flight; six hours in all, and after the debrief, Nick was ready for a weekend. Being an ex-RAF man, he always looked forward to a good meal and evening out on a Friday night. This weekend had to start on Saturday night, but Nick was enthusiastic and ready to go.

By then I was far less jet-lagged than Nick, and it had been an exhausting day for both of us. We went to a restaurant, had a couple of beers, ordered our meal and, just like in a comedy script, Nick was literally falling asleep in his soup. I persuaded him it would be most sensible to give up his Saturday night and reconvene on Sunday. Disappointed but in no state to argue, he agreed.

Lockheed L-1011-500 TriStar pictured over the Lockheed Flight Center, Palmdale, California. (*Lockheed Martin*)

We completed all our flight testing the following week. It was an object lesson from Lockheed in how to approach the business of satisfying the requirements of a foreign authority, without drama or controversy and at minimum cost. They had observed the CAA's approach to certain compliance issues from John Carrodus and Don Burns during earlier certification evaluations. Lockheed had ensured, during our 'training visit', that Nick and I were going to adopt the same interpretations. They had designed, developed and tested the 500 with all this in mind. In Lockheed's ever conscious efforts to save flying time and expense, with the mutual agreement of all concerned, on one flight we flew with the FAA pilot, Carl Jacobson, in the left-hand seat and successfully carried out a combined FAA/CAA assessment of the autopilot performance and failure cases. At the end of our flight tests, we only found one unacceptable issue: when the autopilot was used to execute a 'go-around', it allowed the speed to fall below the minimum scheduled for this situation. This was quite an easy problem to fix and consequently, Lockheed were able to deliver the first 500 TriStar to British Airways on time, one month later.

However, in deciding on the design of the 500, it was beginning to look as if Lockheed had misjudged the market. It was their goal to produce an aeroplane

Lockheed Palmdale flight-line at night, March 1979. (*Author*)

Nick Warner with Lockheed test pilot Don Moor in L-1011-500 TriStar, Palmdale, March 1979. (*Author*)

that would be the first choice of airlines to replace the considerable number of long-range Boeing 707s and McDonnell Douglas DC-8s that were coming to the end of their lives. The earlier TriStar models were designed for shorter-range, high-density routes. The concept of the 500 was to reduce the passenger capacity from the earlier versions to a little more than that of the Boeing 707. Hence, they reduced fuselage length from the previous TriStars, and increased the fuel capacity to achieve the long range. It was becoming apparent that passenger numbers were growing significantly, and the much larger Boeing 747, as well as the long-range version of the McDonnell Douglas DC-10, was taking much of the Boeing 707 market.

During our visit, Lockheed explained their future development plans. In an attempt to make the 500 more competitive and more attractive to airlines, their next version would have a bigger wing-span to improve aerodynamic efficiency and reduce the seat-mile cost. Unfortunately, this would normally necessitate an increase in wing strength and, therefore, more structural weight, thus negating some of the theoretical gain. To avoid the extra structure, Lockheed were installing an innovative 'Active Control System' (ACS). In this system, the ailerons would be programmed to move up together in response to any high wing-loading situation, thus reducing the lift on the outer wing section and keeping the total wing-loading within the structural design limit. They expected to have this new version ready for FAA and CAA assessment within eighteen months.

The next three months were taken up with a splendid variety of routine aeroplane flight-testing and simulator assessments. This included a two-day visit to Schiphol, Amsterdam to finish the Fokker F28-4000 certification and the airworthiness flight test of the first UK 400 for Air Anglia. We also agreed a date with Fokker to return in August to assess the F28-4000 'Category 2' approach guidance performance. There was then the not-so-routine final certification assessment of the aforementioned NDN Firecracker with Darrol. Darrol insisted that I should be the flight test engineer as he knew, from previous experience, that I was the most likely to keep my meals down! We had already covered some of the necessary testing during our earlier 'Special Category' flight test assessment and so we only needed two flights to complete the programme.

As an aerobatic aeroplane, a very thorough examination of the spinning characteristics was necessary. It had to be ensured that there were no configurations or circumstances where it would not be possible to recover from a spin. The manufacturer's test pilot had shown us all his results and we were quite confident that he had covered everything. However, spinning is one of the most dangerous of flight tests and we wore parachutes as a precaution. Our second flight was

Darrol Stinton piloting NDN 1 Firecracker, May 1979. (*Author*)

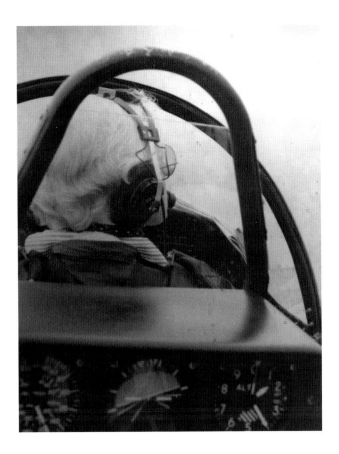

almost entirely devoted to spinning. I had my programme of all configurations to check, each one requiring a spin in both directions and, although there was a specified recovery technique of applying full opposite rudder and some down elevator, we had to be satisfied that inadvertent variations from this technique would not unduly compromise the recovery. For instance, application of aileron in either direction and some variability in the amount of elevator were all checked. The number of turns in the spin and the number of turns in which it took to complete a recovery had to be recorded for each case. In most cases, we let the spin continue for five turns before initiating the recovery but, in a few cases, we waited for nine turns. In all, we accomplished more than thirty spins without encountering any problems. We then had to go through all the standard aerobatic manoeuvres. At the end of all this, Darrol was so 'at one' with the aeroplane that he happily finished off with a demonstration of some of the more complex and esoteric 'tumbling' aerobatics. What a day. Afterwards, Darrol remarked that he did not know how I was able to keep up the enthusiasm for the spinning as he was beginning to feel a bit off-colour!

At the end of the three months, it was back again to Palmdale for the Lockheed L-1011-500 spare-engine pod certification work. Full of confidence in their aeroplane and keen to save expensive flying time, with our full agreement, they again scheduled some combined FAA and CAA certification flights with Nick and Carl Jacobson sharing the piloting. The V_{MCG} assessments, again carried out on the big runway at Edwards Air Force Base, were the most demanding. On the first attempt, even with a 06:00 start, there was too much turbulence to achieve consistent results, so we gave up and continued with other items from the programme. Next day, the turbulence was much reduced, so with FAA and CAA pilots taking turns to check different configurations, the Lockheed scheduled V_{MCG} figures were confirmed. We finished off with two more successful CAA-only flights and said cheerio to Lockheed and Palmdale for what turned out to be more than a year.

Chapter 12

Some Difficult and Rewarding Work

In between several airworthiness flight tests and simulator assessments, Gordon and I returned as planned to Fokker in Amsterdam to make our evaluation of the F28-4000 autopilot Category 2 approach guidance. The F28-4000 had an updated version of the Smiths SEP 6 autopilot, with which, back in my Woodford days, we had achieved good Category 2 guidance results in the HS 748 (see Chapter 6). I had no doubt that Fokker would have been easily able to keep within the Category 2 glide-path maximum deviation criteria with their more stable F28. However, when we gathered for the initial discussions and briefing, it was clear that Fokker were far from happy with the results they were achieving. The results were very inconsistent and, only when they found dubious reasons to discard some individual approaches, were they able to put together a statistically acceptable set of results.

Not filled with confidence, we commenced our flying. The aeroplane was the second 4000 for Air Anglia. Most of the first flight was taken up with our series airworthiness flight test, some remaining aspects of Fokker's production test schedule and Air Anglia's acceptance checks. We terminated the flight with some autopilot approaches into Schiphol which were much as we anticipated from the Fokker analysis. The following day we made two flights dedicated to assessing the Category 2 capability; one at forward and one at aft c.g., making a total of twenty-two approaches. As evidenced from their results, there were extremes of inconsistency and it was clear to us that this could not be considered acceptable. Because this was the aeroplane they were using for their Category 2 development work, it was fully instrumented for the purpose. They were using photographic trace recorders very similar to those that we used at Woodford. After the second flight, I asked if I could look at the instrumentation records.

The following morning, they had the traces ready for us to look at. I had the Fokker engineer explain to me which line on the recorder represented which parameter and he gave me the calibration data for each. Gordon was conducting the debrief, which inevitably included unfavourable conclusions on the Category 2 performance and, at the other end of the conference table, I was able to look at the instrumentation in peace. I managed to correlate the records with my notes from our testing and focused on some of the worst approaches. On the first one I looked at, it seemed to me that the elevator inputs commanded by the autopilot computer were not taking appropriate account of the rate of change in pitch attitude. I looked at some other approaches and the same was true. In fact, it seemed clear to me that when the aeroplane's pitch attitude was moving in a direction to correct a deviation from the glideslope, the autopilot was demanding an elevator input to increase the deviation rather than 'leaving well alone'. I interrupted Gordon's discussion to ask the Fokker engineer if I had correctly understood which direction of pitch-rate trace was a nose-up rate of change and which was nose-down. He confirmed my understanding. By then everyone was looking across the table at me.

I said, with some conviction as well as humility: 'I am fairly sure that, in the autopilot computer, the elevator control law has the plus or minus sign of the pitch-rate input entered the wrong way around.'

Everyone crowded round to have a look and I showed them the cases I had identified. A sort of quiet embarrassment pervaded the air.

Someone from Fokker said: 'We will have to take a look at this.'

Gordon, never one to shower praise around, gave me a look of satisfaction. He wrapped up the debrief and agreed we would await their conclusions. On the way home, Gordon complimented me for finding the fault and we both expressed our amazement that Fokker had not identified the problem themselves.

In my Avro/Hawker Siddeley days at Woodford, our approach had always been that we, the company that designed and manufactured the aeroplane, would always take the lead in any development work, including autopilots, utilizing the full support and advice from the autopilot manufacturer. Fokker, in common with quite a lot of manufacturers, had taken the opposite path and given Smiths the lead responsibility. In some cases, this can be the best approach as the autopilot manufacturer will have a lot of experience in adapting the autopilot control laws to different aeroplane types. The problem can be that the autopilot engineers are fundamentally electronics and computer-minded and not always so familiar with the characteristics of aeroplanes.

One month later, having corrected the error and re-done all their Category 2 approach measurements, we returned to a much happier bunch of Fokker engineers. The results were a huge improvement and, after just one assessment flight, we had no difficulty in expressing our satisfaction. They made no attempt to cover their embarrassment and were full of gratitude for our input in solving their problem.

That evening, Gordon and I flew directly from Amsterdam to Bordeaux for the completion of the CAA Certification of the Dassault Falcon 20F and to carry out the airworthiness flight test of the first UK-imported version.

My old colleagues at Hawker Siddeley, Woodford (now part of British Aerospace) never ceased to continue developing the 748. Rolls-Royce had been able to squeeze yet a little more power out of the Dart turbo-propeller engines and, with some other 'upgrades' including a 2ft increase in wing-span, it was re-designated the Series 2B. Nick and I went to Woodford for our assessment of the 2B. There were a lot of new faces. Tony Blackman had left to become a director on the board of Smiths Industries. The new chief test pilot was Charles Masefield. Although he did not have a military background and had not been to the Test Pilots' School at Boscombe Down, Tony Blackman had seen great potential in him at an early stage and given him more and more responsibility in ongoing flight projects. When Tony was due to leave, he overwhelmingly recommended Charles for the chief test pilot post. At first, this gave the CAA some difficulty. All the UK aircraft design and manufacturing organizations had to be 'CAA Approved Companies'. One important criterion for such an approval was that the senior managers in key positions must have demonstrated appropriate competence and experience in that post. The chief test pilot was one such person. In the recent history of transport aeroplane manufacturers in the UK, there were no chief test pilots who had not been graduates of the Test Pilots' School. Tony Blackman, along with Charles Masefield, had been invited to the CAA to put the case directly to Dave Davies and Gordon Corps. It seems that both Dave and Gordon were convinced. Charles turned out to be an excellent choice and he fulfilled the role in exemplary fashion for many years.

The second surprise for me was to find that the assigned flight test observer was Carol Elliott. When I was at Woodford, she was the 'office girl' in our Flight Test Section. Woodford was obviously somewhere where people's talents were recognized and fostered, irrespective of their background! She performed all her tasks on our flight with an assured professionalism.

As always with the 748, we had the usual list of critical areas in our flight-test programme. However, with some pride, Charles Masefield assured us that the aerodynamic changes had actually improved all the critical stability and control cases. It only took us two short flights to confirm their results and we parted in good spirits.

Following an assortment of airworthiness flight tests and simulators, in November 1979 it was back to the USA for Nick and me. The CAA was receiving a continuous stream of applications for Type Certification of new foreign aeroplane types. Denis Murrin was struggling to fit them all in and, in particular, collect together all the necessary CAA engineers from each department to put together the certification teams. The critical path had become the limited number of test pilots after Dave Davies had retired from flying. Only Gordon and Nick were available for these visits, although Darrol thought he should be considered for some of the smaller types. I gave some support for Darrol in this but, for reasons I never understood, he never succeeded in this quest. Nevertheless, he found himself flying the airworthiness flight tests of some of the larger 'light aeroplanes' in the UK, previously done by Gordon and Nick.

Denis had assembled a team to go to Bethany, near Oklahoma City, where the Rockwell Airplane Company had their facility. The rest of the team comprised Bill Horsley, the usual Vin Wills and Bert Rust, with Mike Benoy evaluating the structures.

The aeroplanes to be the subject of our validation assessment were the Rockwell 695 and 690C models. They were aerodynamically identical, and both had the Garrett TPE331 735 shp turbo-propeller engines, but the version of the engine in the 695 could maintain its maximum power up to a higher altitude. They were both pressurized but the 690C had a 'picture window' in the passenger cabin which, for structural strength reasons, necessitated a lower maximum cabin pressure and, consequently, a reduced maximum cruise altitude limitation compared with the 695. In common with most of its contemporaries, the elevator control circuit needed a down-spring for low speed longitudinal stability and a bob-weight to increase the stick force per g. One issue we found was that, when checking the stall handling characteristics with throttles closed, there was, almost invariably, a significant wing-drop. We soon identified that the cause was a noticeable difference between the idle thrust settings of the two engines. With one propeller generating more air-flow over one wing than the other, the wing behind the lower-powered engine stalled first, causing that wing to drop. We required some improved guidance in the Maintenance Manual to ensure that the two engines had better-matched idle thrust settings.

We did most of our flight-testing on the 695 but finished off by checking all the critical low-speed handling tests with an aft c.g. flight on the 690C. Inevitably, this included the minimum control speeds and associated engine-out handling. For this class of aeroplane, there was no requirement to establish a V_{MCG}. However, the propellers had a reverse thrust setting for use on the ground. Rockwell wanted to take full advantage of this to reduce the scheduled landing distances. There was a potential problem with this. After landing when the pilot selects reverse pitch, there was a possibility that a single failure may cause only one propeller to go into reverse pitch. For this to be acceptable, it was necessary to demonstrate that the

Nick Warner with Rockwell test pilot Paul Leckman in Rockwell 695, Bethany, Oklahoma, November 1979. (*Author*)

aeroplane was controllable after landing with the asymmetric reverse thrust. In common with all previous Rockwell twin-engine types, as we have seen before, it had the unusual nose-wheel steering system with directional control on the ground achieved by differential application of the wheel-brake pads on the rudder pedals. For the shortest possible landing, heavy use of the brakes is necessary but, to retain directional control in this situation, it requires the pilot to ease off slightly on the brake on the side with the propeller in reverse pitch. As soon as a pilot recognizes the reverse failure, cancelling the reverse is the instinctive and recommended action. Before carrying out the last landings of the day to check this case, I recalculated the centre of gravity position. I determined that, with the fuel we had used, the centre of gravity had moved forward during the flight. To regain the aft c.g. necessary for this critical test, I would have to move from my location just behind the pilots to the second row of seats in the cabin.

I gave Nick the speeds and information he needed and added: 'Don't forget, you can cancel the reverse pitch as soon as it becomes difficult to control.'

I know he didn't need me to say this, but I felt better for it. I moved back for the landing. I had quite a good panoramic view from the 'picture windows' on each side of the cabin. On touchdown, Nick selected one propeller to reverse and the aeroplane nose swung from side to side as he tried to control the asymmetry with differential pressure on the brake pedals. He delayed cancelling the reverse for a second or so to allow for a likely pilot delayed reaction. My enduring, vivid recollection of this event has become exaggerated with telling over the years, but I claimed to have been able to see the far end of the runway through each picture window in turn before Nick gained full control! Paul Leckman, the Rockwell test pilot in the right-hand seat, was quite entertained. Having never 'lost' control, Nick was minded to accept it. I never had a convincing explanation of why Rockwell retained this system (rather than the more normal rudder control of the nose-wheel) other than 'it has always been like that' ever since the very first version; the Aero Commander 520 in 1951. However, it seems to be 'second nature' to the pilots who fly them regularly.

As an aside, in the 1960s, the version known as the Shrike Commander was famously displayed by the company test pilot Bob Hoover. He developed a spectacular routine which included single-engine aerobatics; not manoeuvres approved in the Flight Manual. The FAA, after turning a blind eye for quite some time, eventually threatened to revoke his pilot's licence if he continued!

These small, twin-turbo-propeller-powered aeroplanes were always at the most difficult end of the spectrum when it came to assessing the critical flying characteristics. As mentioned before, the amount of thrust and the diameter of the propellers relative to the wing-span were notably more than their piston-engine equivalents. Any situations with asymmetric power or large power changes could be problematic and needed careful assessment against the relevant flight handling requirements.

Additionally, this class of aeroplane was not subjected to the new (at that time) 'safety assessment of systems' requirement in the 'large aeroplane' regulations. In this latter scenario, if a system can fail, its probability of failure must be commensurate with the likely effect of the failure. For instance, in the extreme, if the result of a failure (or combination of failures) is 'catastrophic', then the probability of such a failure must be 'extremely improbable'. 'Extremely improbable' was defined as having a probability of occurrence per flight hour of 1 in 1,000,000,000 (10^9); quite a goal! At the other end of the safety spectrum, a 'minor' failure – one that would require some crew action and a slight reduction in safety margins – must not occur more frequently than once in 10,000 (10^5) flight hours. To demonstrate compliance with all this for every system necessitates many comprehensive safety analysis calculations.

Almost invariably, the small aeroplanes types have much simpler systems. For instance, they have mechanical, manually-operated flying controls. However, it is

Relationship Between Probability and Severity of Failure Condition

Effect on Aeroplane	No effect on operational capabilities or safety	Slight reduction in functional capabilities or safety margins	Significant reduction in functional capabilities or safety margins	Large reduction in functional capabilities or safety margins	Normally with hull loss
Effect on Occupants excluding Flight Crew	Inconvenience	Physical discomfort	Physical distress, possibly including injuries	Serious or fatal injury to a small number of passengers or cabin crew	Multiple fatalities
Effect on Flight Crew	No effect on flight crew	Slight increase in workload	Physical discomfort or a significant increase in workload	Physical distress or excessive workload impairs ability to perform tasks	Fatalities or incapacitation
Allowable Qualitative Probability	No Probability Requirement	<---Probable--->	<----Remote--->	Extremely <----------------> Remote	Extremely Improbable
Allowable Quantitative Probability: Average Probability per Flight Hour on the Order of:	No Probability Requirement	<------------------> <10-3 Note 1	<------------------> <10-5	<------------------> <10-7	<10-9
Classification of Failure Conditions	No Safety Effect	<-----Minor----->	<------Major----->	<--Hazardous-->	Catastrophic

Note 1: A numerical probability range is provided here as a reference. The applicant is not required to perform a quantitative analysis, nor substantiate by such an analysis, that this numerical criteria has been met for Minor Failure Conditions. Current transport category aeroplane products are regarded as meeting this standard simply by using current commonly-accepted industry practice.

Numerical Safety Assessment Chart: Probabilities and Severities of Failures. For example, if a failure or event is categorized as 'hazardous', its frequency of occurrence must be 'extremely remote' (a probability of less than 107 per flight hour; 1 in 10 million). This table is taken from the EASA Certification Standards for Transport Aeroplanes, para 25,1309. It is identical in principle to the earlier requirements in the BCAR and JAR, as well as the FAR in the USA. (See Appendix I)

not unusual to use electrical powered servos for controlling the trim; particularly on the elevator and operating the wing flaps. Just as in the case of simple autopilots, there are several failures that would cause the servo–motor to operate continuously without being commanded; anything from a jammed switch to an electrical failure in the control system. Without having to carry out a detailed failure analysis, such failures must be considered to be 'probable'.

It is necessary, therefore, to assess the consequences of such failures in flight, to demonstrate that the pilot can regain control without endangering the aeroplane. The philosophy is that the pilot would recognize that something was wrong and act to recover the flight path. As in the case of autopilot runaway tests previously described, a delay of two seconds from initiation of the failure to commencing pilot recovery is judged to be a reasonable 'recognition and reaction' time in most scenarios. In the case of an autopilot, the first action will be to disconnect the autopilot after the two seconds. This will immediately cancel the fault and allow the pilot to regain control. With simple electrical trim and flap systems, there is no failure-monitoring system and rarely any means for the pilot to disconnect the servo other than moving the control switch in the opposite direction. Depending on the nature of the failure, this may or may not work and if not, the servo-motor will continue until it reaches full travel. After the two seconds, when taking control, the pilot must be able to overcome the inevitable control forces and recover the aeroplane to a safe situation.

In the case of a wing-flap 'runaway', the control forces should be manageable because the change-of-stick-force-with-flap controllability requirements cover most of the flap angle range. If the flaps are extending, the pilot will reduce the airspeed by reducing power and/or by pulling the nose up to avoid exceeding the flaps-down limit speeds. If the flaps are retracting from fully down, an increase in the power and pushing the nose down will be necessary to increase the airspeed to avoid a stall. Although rarely a serious difficulty, it is often necessary to check these cases in flight.

With an elevator trim 'runaway', the control forces can build up quite quickly and sometimes become unmanageable. If the failure scenario is such that moving the trim control switch in the opposite direction has no effect, an unsafe situation can result. Elevator trim 'runaway' testing is always on the CAA flight test programme for such aeroplanes. If it turns out that a continuous 'runaway' to full travel has potentially hazardous consequences, then some 'instinctive' disconnect switch will need to be incorporated on the trim control selector.

The visit to Rockwell was followed by some interesting and varied activities. Among the usual airworthiness and series flight tests, I had another Concorde flight with Gordon Corps at Filton on the latest production version; always enjoyable and demanding, working to keep up with our test programme as Gordon flew the aeroplane through the flight profile up to the maximum altitude limit of 65,000ft and Mach number limit of 2.12, followed by the descent phase and landing.

At the end of 1979, Andrew McClymont left the Flight Test Performance Section to join Denis Murrin as a design liaison surveyor. Up until that time Denis had been responsible for the evaluations of non-European foreign aeroplanes. The European types had been the responsibility of Austin Kennedy and, after his retirement, the decision was taken to combine all foreign aeroplane evaluations under Denis Murrin. Denis needed an extra person and Andrew was the best choice. Initially, Andrew concentrated on the European types, but later the balance of workload soon led to him dealing with the Chinese Harbin Y-12 and some US types. The CAA Flight Department had sufficient notice of Andrew's move and within a few weeks Graham Skillen, initially from Shorts Aircraft in Belfast but latterly with BAe at Filton, had joined Bill Horsley in the Performance Section.

In early 1980, Nick Warner and I had cause to pay another short visit to Lockheed at Palmdale. Lockheed had reached the same conclusion as McDonnell Douglas: that it was more cost-effective to arrange for the customer acceptance and issue of the UK C of A at their facility in Palmdale rather than in the UK. Although the CAA certification work on the L-1011-200 version of the TriStar

British Airways Lockheed L-1011-200 TriStar, just prior to delivery February 1980. (*Lockheed Martin*)

had been completed about four years earlier, with John Carrodus and Don Burns doing the flight-test work, it was only now that the first 200 aeroplane was being delivered to British Airways. For Nick and me, there were one or two issues outstanding from the certification tests to check and we would be carrying out our series flight test for Certificate of Airworthiness issue. Once our test flights had been satisfactorily accomplished and the CAA had granted the C of A, British Airways needed to have their flight crew on site to go through their acceptance process which included their own flight checks.

Terry Lakin was the British Airways TriStar fleet captain and he had with him Dave Cretney, his technical co-pilot and John Webster, the BA senior flight test engineer. These we knew well from several airworthiness flight tests we had carried out with them in the UK, and Nick and I had arranged to meet up with them in the Wild Goose after our flight to Los Angeles. I knew I would be doing the driving to Palmdale, so I was taking it easy with the beers; not so for Terry and Nick! After a couple of hours, suitable fed with the Wild Goose free pizzas, it was time to leave. Terry wanted a race back to Palmdale; a silly idea in the circumstances, but I agreed. For sensible reasons John Webster decided to come in the car with me and Nick. We made a good start and reached the freeway without any sign of Terry's car. I was quite sure he had not managed to get ahead somewhere in the traffic, but I continued at a speed somewhat greater than the 55 mph limit. We were about twenty minutes from the turn-off to Lancaster when I became aware of the lights of a car behind slowly catching us up. Thinking it was Terry, I decided I was driving as fast as I was prepared to go and if he wanted to pass me, so be it. It was just then that I saw the blue flashing lights on the car behind! I stopped and got out to speak with the Highway Patrol officer. These traffic police were not known for having any sense of humour and he was no exception. He was not impressed with my UK driving licence, nor was he satisfied with my claim that the car was incapable of achieving the speed he claimed I was doing on the uphill road. At this stage, Nick and John had got out of the car to come and see what was happening. His reaction was sudden and positive.

He pointed to our car with his hand by his gun and said: 'There is a car there; go sit in it.'

They returned to the car. At any time, I was expecting Terry to be driving up the road at high speed, which might have taken the heat off me, but it never happened. The officer must have detected that I had been drinking and he gave me his standard tests. They did not have breathalyzers in those days. I was asked to stand on one leg and hold my other foot up. He did the same as a control; I was somewhat steadier than him. He asked me to recite the alphabet backwards; not easy at the best of times, but apart from starting with 'zed' instead of 'zee', he was satisfied after several letters. Because I didn't have a US driving licence, he had no means of directly imposing any penalties. His only option was to complete a form requiring me to appear at a court to answer the speeding offence. I had to sign the form confirming my agreement to this. I protested that I would not be in the USA on the day specified and I could not sign something with which I was unable to comply.

His response was: 'Sign it or you are going to jail right now.' I signed it and he sent me on our way with warnings to keep to the speed limits.

After we reached the Lancaster Motel, there was still no sign of Terry. They arrived much later, and it transpired that Terry's driving was so bad that Dave Cretney had insisted he should take over, which Terry had agreed to, and they continued at a reasonable steady pace. I subsequently wrote a letter to the US Court explaining my situation and I heard no more.

The flight-test programme at Lockheed all went well, and we again successfully carried out a combined CAA and FAA assessment of some of the recent modifications.

Terry Lakin's adventures with his hire car took another turn. One evening he and the BA crew decided to go to a restaurant near Edwards Air Force Base. At some stage, they took a wrong turning and, in the half-light, Terry had not noticed the barbed-wire fence across the road. The lower wire punctured one front tyre and the upper wire gouged serious scratches into the roof paint. They limped back to their hotel with the emergency spare wheel. The story goes that next morning Terry was on the phone to the car rental company complaining that the tyre had blown out, causing the accident with the barbed wire. The car rental company was not ready to be convinced, but when he returned the car the next day he continued with his complaint about their lack of checking on the state of the car's tyres. With a stroke of luck Terry did not deserve, he noticed another hire car ready for collection that had a flat tyre. No further argument was necessary, and his car hire fee was waived.

In Denis Murrin's busy programme, the next aeroplane for a CAA validation assessment was the Piper PA-42 Cheyenne III. This was probably the ultimate turbo-propeller development of a twin-piston-engine aeroplane. Piper had taken the PA-31T development of the piston-engine Navajo, as covered in the previous chapter, and made further significant changes. The Pratt & Whitney PT6A engines were up-rated to 720 shp compared with 620 shp of the PA-31T. The other most noticeable new features were a fuselage extension of about 3ft (1m), the tailplane mounted on top of the fin and a stability augmentation system (SAS). The SAS was an innovation for this class of aeroplane. It was effectively a variable

Piper PA-42 Cheyenne III with Gordon Corps, Lakeland, Florida, March 1980. (*Author*)

down-spring system that used inputs from a wing angle-of-incidence sensor. The elevator control force was gradually increased as the elevator was moved up to reduce speed or down to increase speed. This additional force ensured that adequate longitudinal stability was maintained in all appropriate conditions of power and flap angle at aft centre of gravity.

Piper had moved all their twin-engine aeroplane development and production from their site at Lockhaven to a facility at Lakeland in Florida. As it was a Piper aeroplane, I was back with Gordon Corps. Late March was an ideal time of year for a visit to Florida and the team was grateful to leave the UK winter for several days.

This turned out to be a fascinating investigation. In the UK, the CAA is responsible for granting an 'approval' to the UK company that designs and manufactures the aircraft. This enables the CAA to place reliance on the company to produce 'compliance reports' which the CAA can accept without question if it so chooses. However, the CAA engineers routinely check the company's compliance data in the critical and marginal areas. In the cases of compliance with the flight requirements, in the earlier chapters there are examples of such CAA flight tests on the HS 748 and HS 125. The system in the USA is somewhat different. Instead of approving the aircraft company, the FAA will grant suitably-qualified individual engineers in a company the status of a Designated Engineering Representative (DER). Although these DERs are working for the company, they have a second responsibility to the FAA, which enables them to directly assess the means of compliance with the requirements within their engineering expertise. They are then able to sign off the company compliance reports on behalf of the FAA. However, the FAA engineers in the local regional office are, themselves, always at liberty to check the DERs' findings in critical areas. In practice, particularly with the smaller aeroplane types, the Regional FAA engineers have a close working understanding of what constitutes an acceptable means of compliance with the DERs operating in their area which, with the FAA and the company working to the same criteria, tends to ease the development and certification process.

However, the Central FAA Office in Washington is always at liberty to involve itself in the certification process. In this case of the PA-42, they thought the latest changes may be of such significance that they were outside the normal experience of the local FAA engineers in the Regional Office in Miami; they may not be able to judge reliably the validity of the Piper DER's findings. Some FAA engineers from the Washington Office were assigned to support those in the Miami Office. To some extent, we in the CAA, having carried out our validation assessments of various aeroplane types from most FAA regions, had become quite well 'calibrated' with the standards generally accepted by most FAA regions, including some of the 'subtle' inter-regional differences.

Consequently, we were confronted with some unusual compliance decisions in certain areas, the most significant of which was the addition of a 'stall protection system'. The FAA pilot, who it seemed had little experience of current twin-turbo-propeller types, considered that the stall characteristics with power on were unacceptable. The nose-drop was not very pronounced and there was a tendency for one wing to drop. As a consequence, Piper had to install a system of warning lights and a horn to help the pilot identify the point of the stall. It was our conclusion that the PA-42 stall characteristics were, in fact, a slight improvement on some of the earlier Piper PA-31 variants and no worse than other contemporary twin-turbo-propeller types we and the FAA had previously certificated. We concluded that these warning lights and horn were an unnecessary distraction and were not to be fitted for UK certification. Furthermore, our Bert Rust was a bit concerned about the lack of integrity of this system and the possibility of false warnings, so our decision that it should be removed relieved Piper of this potential headache. We found a few other problems, but nothing insurmountable for UK Type Certification.

This visit to Piper was closely followed by a week and a half back with Dassault at Istres in the south of France, again with Gordon Corps. This time it was their

Dassault Falcon 50. (*ABPic, Samuel Parmentier*)

top-of-the-range corporate jet, the Falcon 50. Developed from the Falcon 20, it had a lengthened fuselage, a completely redesigned wing and three engines in the tail like the HS Trident, the Boeing 727 and the TriStar arrangement with an 'S'-duct air intake to the central engine. The engines were Garrett TFE731-3s; it had a maximum airspeed of 484 knots, a cruise Mach number of 0.80 IMN, an impressive maximum altitude of 49,000ft and a range of 3,000 nautical miles. It had accommodation for two crew and up to nine passengers.

We did most of our flying on the prototype aeroplane at Istres. However, for some of the system failure testing and evaluation of the production flight-deck, we went to the Dassault production facility at Bordeaux to fly one of their pre-delivery aeroplanes. We were fortunate to be there over the middle weekend of our programme and were treated to a visit to the Dassault family vineyard which, of course, included quite a bit of 'tasting'. I had no idea that one of the surviving members of the Dassault family owned a vineyard and he personally showed us around. Gordon naturally purchased several boxes, but I settled for just one box of mixed vintages. I was advised to wait a few years before opening the bottles; good advice which I followed, and all the bottles were a pleasure to drink.

There were two unexpected problems resulting from our tests. For some reason, the stick force per g was much lower than both the BCAR and FAR requirement. When confronted with this, Dassault were a little surprised but, with his good understanding of French, Gordon formed the impression that they had not fully investigated the case. An aeroplane with powered flying controls needs to have an artificial feel system (an electro-mechanical system to generate control forces) and it is quite usual to tailor such a system to meet the control force requirements in all conditions. Furthermore, the CAA systems engineers found that there were single failure cases in the feel system that would reduce the stick force per g to less than 5lb elevator control force to exceed the 2.5 g wing strength limit. Such an extremely low force could easily be applied inadvertently; this could potentially be catastrophic, although the actual g at which the wing would break off is quite a bit more than the 2.5 g. The second significant issue from the CAA certification required fitting a stick-shaker to warn of the approach to a stall. Modern aeroplane wing design had largely eliminated the characteristic of natural aerodynamic buffet just prior to reaching the stall. Consequently, it was the universally accepted solution to fit a stick-shaker, which operated at an appropriate speed above the stall speed, thus replicating the warning of the pre-stall buffet. On the Falcon 50, there was natural buffet in some configurations, but it was not sufficiently attention-getting, nor did it always give a sufficient margin before the stall. Dassault had taken the view that supplementing this low and variable buffet with a warning horn would suffice. This was a solution generally accepted on the smaller BCAR Section 'K' and FAR Part 23 aeroplanes, but the CAA never considered it to be an acceptable option for the BCAR Section 'D' large transport aeroplanes.

Three months later, in August 1980, Nick Warner and I were back in Palmdale to make our long-awaited flight assessment of the Lockheed TriStar 500 with the Active Control System (ACS). Denis Murrin, with some of the CAA systems and structures specialists, was also at the Lockheed main design and manufacturing plant at Burbank in north Los Angeles to complete their assessments.

After the usual friendly greetings, we started into the first briefing session. It was clear that Sam Wyrick was not his usual ebullient self. With support from his engineers, he explained that they had encountered some problems during pitching manoeuvres at high Mach numbers. When manoeuvring in the nose-up direction, the ACS moved the ailerons up together to reduce the aerodynamic load on the outer wing which, because of the swept wing, moved the centre of lift forward and caused the aeroplane to pitch up. This effect had obviously been anticipated and generally had only a small effect. However, it transpired that the consequences at the higher Mach numbers were much more significant than expected. Lockheed had done a lot of testing and analysis and formulated a relationship between Mach number, aeroplane weight and air density (a function of altitude) at which the pitch-up problem was at its worst. They concluded that the Mach number was the prominent factor and had carried out most of their exploratory testing at the normal maximum operating Mach number limit and at critical weight and altitude combinations. The rate of pitch-up could be very abrupt in the worst cases and, if the pilot did not react rapidly, the aeroplane would reach a level of g that could exceed the wing strength. In these situations, there was no aerodynamic buffet to warn or deter the pilot, which was often the case when manoeuvring at high Mach numbers. As an alternative, to provide the pilot with some warning of the approaching dangerous situation, Lockheed had added a modification to operate the stick-shaker at 1.75 g.

The FAA had made their assessment and concluded that they could accept this characteristic because of the unlikely combination of flight conditions in which it could occur, coupled with the extra deterrent of the 1.75 g stick-shaker.

To investigate this, the test technique was to steadily and carefully increase the normal acceleration (g) in a banked turn, increasing the angle of bank gradually to increase the g. As the g approached 2.0 (60 degrees of bank), the aeroplane pitched up suddenly and the pilot had to take instant recovery action to prevent reaching the structural limit of 2.5 g (as mentioned before, there was always a margin between this limit and the point at which the wing would fail). Obviously, the critical situation was with the aeroplane loaded at the aft c.g. limit. We made our first flight in this configuration to get an early opinion of this problem. We

Nick Warner with Lockheed test pilot Bill Weaver in L-1011–500 TriStar; Sam Wyrick on extreme left, August 1980. (*Author*)

Lockheed L-1011-500 TriStar in a 65 degrees banked turn over the Rockies, assessing the 'pitch-up' problem, August 1980. (*Author*)

were able to accomplish quite a lot of the other aft c.g. testing on this flight, but the pitch-up characteristics were top of the list. The piloting skill involved in slowly winding the aeroplane into an ever-tightening turn while maintaining the speed at a constant Mach number close to the maximum limit is extremely demanding. Nick was fantastic. The saving grace in this situation is that the pilot knows the problem is coming and is poised for an instant recovery action. The degree of the sudden pitch-up we found was exactly as we had seen in the Lockheed test results. Conducting 60 degrees plus banked turns with a back-drop view through the cockpit window of the Rocky Mountains remains a vivid memory.

In the debrief after the flight, there was only one question. What was our conclusion? Lockheed needed an answer and we needed to consider our decision. We knew that the FAA, after some deliberation themselves, had decided to accept it. Their decision was largely based on a safety analysis probability argument, with credit for the 1.75 g stick-shaker. Firstly, the aeroplane had to be flying at or close to the maximum Mach number limit and then for some reason, the pilot needed to initiate a rapid manoeuvre that would increase the normal acceleration to around 2.0 g. For the aeroplane to be flying at the maximum Mach number, which is significantly higher than the normal cruise Mach number, requires some unusual event to have occurred; a very significant clear air wind-shear or a failure of the Mach number indicating system causing the pilot to inadvertently increase the actual Mach number. These possibilities are, in themselves, fairly remote. Add to this the unlikely need for the pilot to be making a relatively violent manoeuvre, and the overall probability becomes more remote. Given that the pilot might react rapidly enough to limit the pitch-up, as well as the fact that there is a structural strength safety margin above 2.5 g, the event itself, although certainly 'hazardous', may not always be 'catastrophic'. With all this in mind, I collected a selection of the instrumentation records and a copy of the pitch-up flight-conditions criteria chart, and Nick and I returned to the hotel.

We certainly felt that the decision was going to be controversial and crucial for Lockheed. We decided, after we had collected our thoughts, that we should make a phone call to the CAA first thing next morning (with the eight hours' time difference, that would be late afternoon in the UK) to apprise our Flight Department colleagues of the situation. We would not expect an instant response, but we could anticipate feedback within a day or so. Having chatted over dinner, Nick was erring towards the same conclusion as the FAA. I was inclined to agree, but there was something in the back of my mind about the Lockheed chart that was worrying me. It was not a straightforward chart to interpret, having to show the relationship between three variables: the degree of pitch-up, the Mach number and the value of the weight/relative density ratio. The way it was drawn tended to emphasize the Mach number effect.

Back in the hotel, I pondered over the chart and did some calculations with my Casio. It then became apparent that the Mach number was not quite as dominant a factor as we were all convinced. I found that at lower Mach numbers, around the high end of the typical cruise Mach number range and at a particular combination of aeroplane weight and air density, the pitch-up characteristic looked to be more severe than the cases we had been testing. To add to the seriousness of this new situation, the worst-case combination of aeroplane weight and altitude was an entirely possible scenario in the cruise. Nick checked my calculations and was equally amazed. There was no point in making the CAA phone call in the morning; we first had to check out this new conclusion with Lockheed.

In the morning meeting, Sam Wyrick and his engineers were cautiously anticipating our provisional conclusions. They were not expecting the bombshell we presented to them. They listened to my conclusions and checked the calculations. It was clear that some more flight-testing was required to confirm the validity of their chart in this new combination of flight conditions. It would be usual for the manufacturer to thoroughly assess all flight characteristics and satisfy themselves before inviting the regulatory authorities to fly the aeroplane. Some discussion took place. To Nick's credit, the Lockheed test pilots, supported by Sam Wyrick, had absolute confidence in Nick's capability and expressed their willingness for him to do the flying to save the need for duplicating the flight tests. There was clearly some significant risk involved and Nick's agreement was paramount, as was mine if I was to be participating in the flight. Clearly, this option was putting us (the Airworthiness Authority) in the position of conducting the manufacturer's testing. There were safety considerations. We had a little chat. Nick was ready to go ahead. I was very confident in Nick's capability for this demanding task and, if we declined, we would still have to repeat the critical tests after Lockheed had completed their flight testing; it was not a difficult decision for me.

Together with Lockheed we put together a flight plan to investigate the new problem configurations, obviously starting with less critical cases and building up to the potentially worst case. Later that day, we set off. It was soon clear that the results were following the predictions of their chart. We were looking at pitch-up situations somewhat worse than those previously experienced. For Nick, the task of setting up the configuration and increasing the g in a banked turn was slightly less demanding because we were at lower Mach numbers than the maximum Mach limit as in the previous cases. However, the suddenness of the pitch-up necessitated an extremely rapid reaction on Nick's part to recover. We progressed slowly to the worst case and as soon as it was clear that the new predictions were valid, we called it a day. It was certainly not a scenario to be delving into any further in the search for academic data. It was an extremely dangerous test that could result in a catastrophic event. In some of the critical situations, the aeroplane was beginning to wind up into the pitch-up at or slightly before the onset of the 1.75 g stick-shaker.

In the following debrief, it was no surprise to anyone that we found this unacceptable. We were no longer just considering a high Mach number event and so the probabilities were much more likely. Additionally, the severity of the pitch-up was greater. Lockheed had an idea up their sleeve. They had not been totally sure we would accept the original situation and had been developing a means of alleviating the pitch-up. In addition to the active ailerons to relieve the wing loads, there was also a Direct Lift Control (DLC) system. This was designed to respond to sudden changes in wing lift during turbulence. A series of hinged spoiler flaps on the upper wing surface would rise up to dissipate some of the wing lift. The Lockheed proposal was to programme these spoilers to operate at the point of the pitch-up, to reduce the total lift capability of the wing and hence limit the amount of g generated. In view of our finding the different critical scenarios, Lockheed would need to do some re-programming of this system, which they referred to as Manoeuvring Direct Lift Control (MDLC). While they worked on this revision to the MDLC, over the next few days we continued with the remainder of our certification flight-test programme without problems.

So, for our final flight, we were back over the Rockies repeating the high g manoeuvres. There was no doubt that the MDLC had improved the severity of the g peak, but not to a degree that we felt would be acceptable.

At the debrief, I made a throwaway remark: 'The MDLC is helping but what we need to do is destroy much more of the wing lift so that it cannot generate a structurally dangerous situation,' and then I added, without giving it much thought: 'What about the speed brake system?'

The speed brakes are large panels on the upper surface of the wing that extend almost vertically to significantly increase the aeroplane drag and greatly reduce the wing lift. As well as being used in flight to reduce airspeed, they are also deployed on landing to reduce landing distances. We left Lockheed to ponder this possibility, without any idea whether the TriStar 500 would ever gain a UK Type Certificate; a critical situation as British Airways had placed firm orders for the type.

Back in Redhill there had to be a serious discussion of our recent results. Were we going too far in requiring further improvement? It all came down to probabilities and consequences. What was the probability of the aeroplane being in the critical flight conditions of weight, altitude and Mach number and what was the probability of the pilot needing to execute a serious nose-up recovery or avoidance manoeuvre? If this combination of events did occur, then what was the resulting effect likely to be? Initially, we looked at the set of conditions we first assessed before the realization that there was a more critical scenario. The consensus was that the probability of being in the high Mach number flight condition and needing to execute the rapid nose-up manoeuvre was extremely remote. If this situation did occur, it was not certain that the consequences would be catastrophic if the pilot took brisk recovery action, encouraged in part by responding to the 1.75 g stick-shaker. This combination of probability and consequences was such that the likelihood of a catastrophic outcome was at, or beyond, the safety goal of 1 in 1,000 million flying hours and therefore acceptable.

However, it was clear that the more critical cases at lower Mach numbers, having both a higher probability of occurrence and a more rapidly developing pitch-up, were no longer within the acceptable criteria for the probability of a catastrophic event. It was also clear that, in some critical cases, the operation of the 1.75 g stick-shaker was too late to greatly influence the pilot recovery action. The improvements effected by the MDLC were carefully assessed. The resulting consensus supported the conclusion reached by Nick and me, that the MDLC improvements were not sufficient to achieve an acceptable situation.

Six weeks later, we were back at Lockheed. They had made some refinement to the MDLC that included adding an input to deploy the speed brakes to reduce the wing lift in the critical cases. This they referred to as the Recovery Speed

Brake (RSB) system. They had also added some vortex generators on the upper surface of the wings that re-energized the local airflow and reduced the Mach number compressibility effect. From their own testing, they had shown that this combination had significantly improved the critical cases and were fairly confident that we would now find it to be acceptable. We looked at their test results and had no doubt that things were much better.

We enquired how the FAA had responded to the new findings. They had recognized that the conditions for the event were somewhat more likely and, without the MDLC and RSB, the pitch-up was more severe. However, they retained their confidence in the deterrence of the 1.75 g stick-shaker and they considered that the probability of encountering the situation was still sufficiently remote not to require the MDLC/RSB modification.

Early next morning, with the aeroplane loaded to aft c.g., we took off with a degree of confidence. As before, we did not start at the most critical conditions, but the improvement was immediately clear. The degree of pitch-up had been marginally improved by the modifications to the MDLC, but the operation of the RSB immediately dissipated much of the wing lift which significantly reduced the maximum g-force that could be generated. We repeated the test in various configurations, making several repeats at the most critical case. Knowing that the operation of the RSB limited the wing lifting capability, which greatly reduced the probability of reaching the limiting wing strength, made it a somewhat less demanding test to carry out. When the pitch-up developed, it was not so essential to take the recovery action as rapidly as before and therefore, the consequences of a more measured reaction to the event could be assessed.

Before the debrief after the flight, Nick and I had a private chat. It was immediately clear that we had both reached the same conclusion. The event was no longer potentially catastrophic and arguably not hazardous. Although the worst cases had now been found to be within the possible range of normal operating conditions, the probability of the pilot needing to execute such a nose-up manoeuvre was still unlikely. It would be our recommendation to accept it, but we would need to take our conclusion and the evidence back to Redhill for a thorough review within the CAA by all concerned. We put this to Lockheed at the debrief and Sam Wyrick, their pilots and engineers were visibly more cheerful than we had seen for a long time. Sam seemed to be optimistic that we would be able to convince the CAA back home.

There was still more flying for us to do in the next few days. Apart from some areas we had decided not to complete last time until we were sure we had a potentially certifiable aeroplane, there were several aspects, particularly manoeuvrability and stalling characteristics that could be affected by the operation of the MDLC system in other flight configurations. Normal stall handling and stalling speeds were not affected but, as anticipated, the MDLC was having some input during the dynamic and turning stalls; to everyone's relief, this did not significantly impinge on the stall handling characteristics. Stick force per g was another area affected by the MDLC, but the minimum requirement of 50lb to 2.5 g was still met.

Back in Redhill, we explained the significant modifications Lockheed had made to remove our concerns, noting that the FAA had concluded they were not necessary for a USA Type Certificate. It was always very disappointing to require the inconvenience and expense of additional modifications to be fitted for UK Certification. Were the safety concerns we had identified sufficient to justify such modifications? When making safety assessments of systems, it is usually relatively straightforward. The extensive testing of equipment components gives good information of their likely failure rate (defined as a mean time between failures: MTBF) and the consequences of the failure will usually be understood. See the Safety Assessment requirement chart earlier in this chapter. In the case of in-flight situations, the probability of a certain combination of flight conditions arising is more subjective, as is the likely pilot reaction. The consequences, however, may be more definable. If the consequences, with all the necessary combinations of adverse circumstances and pilot responses, are catastrophic, then it must be certain that the probability of the occurrence of this combination is extremely improbable. It was the judgement of the CAA that, without the MDLC and RSB modifications, the consequences could be catastrophic. Was the combination of circumstances and actions necessary to reach this catastrophic situation better or worse than extremely improbable? Here the potential size of the whole fleet of the aeroplane type can be significant. For example, if there are 1,000 of such aeroplanes, then the probability of any event occurring in the lifetime of the fleet is 1,000 times greater than if there is only one of the type! This can be significant. How many L-1011-500 aeroplanes could be expected to be in the operating fleets of airlines under the jurisdiction of the CAA? The CAA needed to err on the side of safety and conclude that the MDLC and RSB were necessary.

Sadly for Lockheed, the L-1011-500 was the wrong concept for the time. Passenger numbers were increasing rapidly, and operators needed aeroplanes with more seating capacity. Boeing and McDonnell Douglas were stretching the size of their types, not shrinking them. In the end, there were only eleven L-1011-500s on the combined registry of the UK and UK-dependent countries. In total, there were only about fifty worldwide. With the benefit of many years of hindsight, it is the author's view that, in this case, we in the CAA may have been over-cautious and the FAA had possibly made the correct call.

Chapter 13

Time to Move On? Tragic Accidents

I n mid-1980, Gordon Corps was our next pilot to be tempted away from the CAA. As our project pilot for the Airbus aeroplanes, he was very well-known and respected by Airbus Industrie and when they were looking for a new test pilot, Gordon was their prime potential candidate. The CAA test pilots were quite well paid in comparison with UK manufacturer's test pilots and senior airline pilots, but the salary offered by Airbus Industrie was very much greater. He spoke French well and it was not a difficult decision for him to accept and move with his wife to the Toulouse area where Airbus had their factory and Flight Test Centre.

Gordon was able to give the CAA plenty of notice which allowed time for a new pilot to be recruited. At that time, Head of the Flight Department Ronald Ashford was transferred to be Head of the Airworthiness Division Administration Department and Dave Cummings was promoted to take his position as Head of the Flight Department. Keith Perrin would have been the obvious first choice, but Keith decided that, from his perspective, being chief flight test engineer brought with it a greater level of job satisfaction. Dave Cummings' move left another vacancy to be filled, for a flight test engineer.

In a surprisingly short time, the department had recruited a new test pilot, Al Greer, and a flight test engineer, Dave Morgan. They were both previously working at the Aircraft and Armament Experimental Establishment (A&AEE) at Boscombe Down. Al was a graduate of the Test Pilot School and had continued at the A&AEE, test-flying new military types for the MoD. Dave was one of the A&AEE permanent flight test engineers, just as John Denning had been before he joined the CAA. They both needed time to assimilate the fundamentals of their tasks as Nick Warner and I had several years earlier.

A year or so before, I had given some thought to the prospects for promotion within the CAA. Without really having any clear idea of what path I might take, I had worked out how many promotional steps there were to reach a department head level; a reasonable goal on which to set my sights, I thought, even if it became just a dream. It was normal, if one made good progress in one's early career, to be 'promoted in post' to the next level and I had achieved this in 1980. This would leave three more steps to reach a department head level. The retirement age was sixty and that gave me about another twenty years to fill. It was not 'rocket science' to realize that, to give me a useful number of years in each higher post, I needed to be moving on in the next year or two; about seven or eight years from joining the CAA Flight Department.

The obvious next step would be to follow in Keith's shoes, but I could see no sign of Keith having any reason to vacate his post in the foreseeable future. Just like Dave Davies, Keith was a fantastic role model and was totally on top of his game. Did I want to emulate him? Did I want to carry on until he felt he needed a change himself? There was no shortage of interesting new aeroplane types on the horizon and no chance of being bored with the job. However, I was feeling that, no matter how interesting and demanding the future flight-testing might be, it would be, to some extent, more of the same.

Earlier in March that year, while carrying out the flight-testing of the Piper PA-42 in Lakeland, Florida, Denis Murrin quietly asked me in the bar one evening if I would be interested in joining him as a design liaison surveyor in the Foreign Aircraft Section. Of all the career paths I could possibly consider, this appealed to me much more than any other and I did express an enthusiastic interest. The task of a design liaison surveyor was to manage the teams of CAA specialist design surveyors (including the flight test team) who would be undertaking the investigation of foreign aeroplane types for the ultimate goal of granting a UK (CAA) Type Certificate. Although Andrew McClymont had left the Flight Department a year or so earlier to become Denis's assistant, primarily on European aeroplane types, the number and variety of applications from foreign manufacturers seeking UK certification was increasing rapidly. Denis was no longer able to manage all these, even with Andrew's help, and he was trying to gain approval to take on another surveyor.

In the subsequent three or four months, I heard nothing more and life continued with plenty to keep me busy. However, in July Denis had received the approval he needed to employ another design liaison surveyor and a notice inviting applications was published in September. Denis encouraged me to apply and I had no hesitation in taking his advice. Things rarely moved very quickly in the CAA's recruitment process and it was October, after our second L-1011-500ACS visit to Lockheed, that the short-list of three applicants, me included, were called for interview. Looking at the final list of three, it seemed to me that there was one other candidate, Mike Benoy from the Structures Section, who I thought would be serious opposition. Despite Denis's previous encouragement, this put a little dampener on my enthusiasm and confidence.

On the day allocated for the interviews, the other two candidates were scheduled for the morning and early afternoon, with me later in the afternoon. By chance that day, most of the test pilots and flight test engineers were in the office and it was the first opportunity to get Gordon Corps, Keith Perrin and the others together for a pub lunch to go over our L-1011-500ACS results. This proved to be more time-consuming than anticipated and in my enthusiasm for getting all the details of the flight test results across, I consumed a few beers.

At some point, Graham Skillen said to me: 'Aren't you supposed to be having an interview this afternoon?'

This brought me down to earth with a bump, and we continued for a little while longer with no more beer on my part!

I turned up on time for the interview. Given my previous dubious track record in interviews, my post-lunchtime state may have worked in my favour. I was full of enthusiasm, confidently answering the questions, albeit almost certainly emanating evidence of a drink or two. There were three people doing the interviewing: Denis Murrin, Ronald Ashford (who at that time had moved to be Head of Administration) and a personnel department representative. Denis took the situation in his stride with a mixture of enthusiasm and amusement. Ronald was very formal. He and Denis were quite different characters and they did not always see eye to eye. Typical of Ronald's usual analytical approach to things, he had a prepared sheet in front of him with several headings and a space under each category for writing notes and entering a score. From time to time, I glanced towards his sheet but, although I had become very accomplished at reading upside-down, he kept it well protected. I was very sure that Ronald would have not been impressed with my demeanour and, on most of his criteria, would probably have scored me lower than the other candidates. I envisaged there might be some confrontation of opinion between Denis and Ronald, with the personnel representative possibly having a casting vote.

Soon after the interview and about one month after flight-testing the L-1011-500 with the recovery speed brake modification, we returned to Palmdale in November 1980 to carry out the necessary assessments of the additional configurations that Lockheed wished to include in the CAA certification.

Firstly, we had to complete the V_{MCG} tests that were not done during the previous visit because there had been too much wind. Nick performed these with his usual precision, including a couple of repeat tests to show that the resulting V_{MCG} could be achieved consistently. Interestingly, one of the L-1011-500 customers, British West Indian Airways (BWIA) had realized that, when operating on some of their shorter runways, the V_{MCG} speed was the compromising factor in determining the scheduled take-off distances. Lockheed were in the middle of developing a larger rudder to improve this situation and as BWIA based their certification on the UK CAA standards, we would be required to assess this at a later date.

Secondly, Rolls-Royce had made some changes to improve the fuel efficiency of the RB211 engines. It was thought that these changes may have a potentially adverse effect on the engine acceleration characteristics and the in-flight re-start capability. Lockheed had thoroughly run through a series of tests to check these aspects and were sufficiently confident to schedule a single flight for a combined CAA and FAA evaluation. With me determining the appropriate critical configurations, airspeeds and altitudes in accordance with the agreed flight plan, Nick and the FAA L-1011 project pilot Carl Jacobson shared the flying without the need to repeat any of the tests. Everything was achieved to the satisfaction of all and Carl Jacobson happily went away with a copy of my flight notes.

Next on the agenda was the spare engine pod certification testing. The TriStar had the same problem as the DC-10 (Chapter 10); the fuselage was not big enough to accommodate a spare engine if a replacement was needed down-route. Interestingly, Lockheed found that the aerodynamic drag of letting the air flow through the engine was little different from fitting a streamlined cowl over the intake as McDonnell Douglas had done on the DC-10. Dispensing with the weight and inconvenience of this cowling was a significant advantage.

We were able to assess all the critical aspects of this configuration in two long flights totalling about eight hours. It was important to review the prescribed procedures

Lockheed L-1011-500 TriStar with spare engine pod, Palmdale, November 1980. (*Author*)

for fuel tank usage throughout a flight to maintain a reasonable lateral balance. From this information, we could assess the worst-case scenarios and build them into our test programme. Rates of roll, lateral stability, minimum control speeds and, in particular, stall handling could all be affected by the aerodynamic and the lateral weight imbalance. In normal flight, some imbalance can be compensated, by application of aileron trim, to maintain wings-level flight. However, the resulting amount of applied aileron causes an aerodynamic imbalance which can affect the results of some of the above certification test cases. The tendency for a wing to drop at the stall on the L-1011 was always a critical issue, particularly the 500 with the Active Control System and, with the combination of aerodynamic and inertia effects of the pod, it was necessary to control the sideslip carefully down to the stall to ensure the tendency for a wing to drop was maintained within the 30 degrees bank angle limit. Similar care was necessary during the turning stall tests to restrict the bank angle to within the limit of 60 degrees when starting from a 30 degrees banked turn towards the engine pod. This was not too different from the normal aeroplane configuration and the stall handling was accepted.

Surprisingly, I did not hear any follow-up from the interview until after returning from this visit to Lockheed in late November. Unbeknown to me, at about that time Mike Benoy, the candidate who I felt sure would be the greatest competition, had also applied for a different post as technical assistant to the Head of Airworthiness Division John Chaplin; a post he preferred and was subsequently awarded. I received a letter offering me the design liaison surveyor post. I never knew if I would have been the successful candidate in any event. Without hesitation, I formally accepted the post. Subsequently, Denis made it clear that he was very happy to have me join him, but he never let on any details of the decision-making process.

Although it would be several months before I could take up the new post, there was, consequently, an urgent need to do some reallocation of the flight test engineer responsibilities. The most immediate decision needed to be who would take over from me on the Lockheed L-1011 TriStar projects. Additionally, another flight test engineer needed to be recruited and there would be the usual advertisement in *Flight Magazine*. Keith and I went over the likely candidates we knew of in the UK aviation industry. I mentioned Geoff Stilgoe from Woodford, who I was sure would be very suitable. There were a few other names in the frame, but Keith encouraged me to give Geoff some prior notice to look out for the advertisement, just as Bill Horsley had done for me before my applications. This I did, and Geoff seemed enthusiastic about the idea.

One month later, Nick and I were back again at Palmdale together with Dave Morgan who was to replace me on the L-1011 projects. Whereas I had had the benefit of a tailor-made Lockheed training course, for Dave this was a case of jumping in at the deep end. He would have all Nick's experience to draw on, but he needed to pick up the pieces as comprehensively as possible during this relatively short visit.

Lockheed had completed their certification testing of the big rudder modification. Additionally, the first 500 to the UK standard was ready for delivery to British Airways. Bill Horsley also came along to go over the V_{MCG} results as we knew the big rudder modification was not of direct interest to the FAA, but although the FAA was the certificating authority for all Lockheed's design changes, they seemed ready to formally endorse our results.

Lockheed had scheduled us for a first flight at 6:00 on the Monday morning to start with the big rudder V_{MCG} tests. This meant we would need to arrive at Palmdale on the Thursday to allow time to review their results, prepare and brief our flight plan as well as introducing Dave Morgan to all concerned. We expected this would take up the whole of Friday and possibly some of Saturday morning.

This time, I had a plan for the four of us to do a bit of tourism. The 'Smith Tour', as it jokingly became known, was a visit to Death Valley. Spanning the borders of California and Nevada, it is on the edge of the Mojave Desert and a reasonable driving distance from Palmdale. At its lowest point, it is some 280ft below sea level and, as its name implies, it was not an area in which to be stranded. Being December, the peak daytime temperatures were usually in the mid-20s compared with typically up to 50 degrees C in the summer. It is claimed to have the highest temperature ever recorded on the Earth's surface of 56.7 degrees C. There is a famous hotel at its centre, the Furnace Creek Inn, at which I made reservations for the Saturday night. The scenery is incredible and the history fascinating; well worth a visit for anyone in the region.

Unfortunately, the work on the flight plan that had to be accomplished at Lockheed before we could leave stretched well into Saturday morning. Consequently, with the early sunset at that time of year, it was dark before we arrived at the inn, and we missed seeing much of the desert scenery. 'Smith Tours' came in for much ribbing because of this. Nevertheless, the next day we had a great experience and arrived back at our motel near Palmdale late on Sunday evening. It had been a perfect opportunity for Dave to get to know Nick and Bill.

At 06:30 on Monday we were airborne to commence the V_{MCG} tests, exactly as Lockheed had planned. I was happy to let Dave run the programme in accordance with the briefed flight plan. I looked over his shoulder occasionally without needing to intrude and all moved along as planned. The V_{MCG} speeds achieved in all take-off configurations were some 4 to 5 knots lower than for the normal rudder. This

'Smith Tour' of Death Valley, California, December 1980: Nick Warner, Bill Horsley and Dave Morgan. (*Author*)

was exactly as Lockheed had determined; giving a worthwhile improvement on take-off distances at lower weights for the benefit of British West Indian Airlines.

Although my flight-test programmes were beginning to run down, there was one last visit to Embraer in Brazil. This was to assess the final production version of the Xingu, the EMB 121B, and an increase in the maximum take-off weight on the EMB 110P1 Bandeirante. There were no new Embraer aeroplanes ready for certification until their 'commuter' aeroplane, the EMB 120 Brasilia, in several years' time. Therefore, it was sensible for Nick and me to complete these EMB 121B and 110P1 programmes, leaving decisions on the future allocation of Embraer flight test responsibilities to be decided much later.

The 121B had a slightly longer fuselage. Embraer found that this had removed the 'deep-stall' situation so the stick-pusher was no longer required. To improve the natural stall handling characteristics, they had fitted the ubiquitous 'stall strips' on the wing leading edge. Still being somewhat cautious about the potential for a deep-stall, they had left the stall recovery tail-parachute fitted for our flight tests, a similar arrangement to the one fitted for the HS 125 testing with John Carrodus (Chapter 9). The stick force per g was improved and the bob-weight was no longer

The CAA team en route from Rio to São José dos Campos, February 1981. Front row: Bill Horsley, Bert Rust and Peter Richards; second row: Denis Murrin, Vin Wills and Nick Warner. (*Author*)

fitted, but the longitudinal stability down-spring was still necessary. The stall handling was good, and the critical stability and control cases were acceptable. We came away with only two unacceptable items: the scheduled minimum approach speed had to be increased by 2 knots to meet our minimum rate of roll requirement with one engine inoperative and, for some reason, there was far too much residual thrust with idle power so that continuous use of the wheel-brakes was necessary during taxiing.

The only problem we found with the increased weight 110P1 was a difficulty in achieving the scheduled engine-out climb performance at the higher weight. The atmospheric instabilities in the tropics can often cause difficulties in achieving a steady five minutes for the climb measurement. We left Embraer with the task of producing more convincing data or making a correction to the scheduled performance.

At the end of our flying, I had another opportunity to go back-seat flying; this time in the new Embraer Tucano turbo-prop training aeroplane, with the

Embraer EMB 121B Xingu with Nick Warner and Embraer test pilot Sergio, São José dos Campos, February 1981. (*Author*)

The author after a flight in an Embraer Tucano, February 1981. (*Author*)

inevitable aerobatics and opportunity to fly it myself, including a couple of loops. A modified version of this aeroplane was later chosen by the RAF for pilot training.

With some sadness, we bade our farewells to Embraer. I expected it to be my last time but, in fact, it would not be too many years before I was back in my new capacity as a design liaison surveyor. It was certainly the last time I would fly with Nick; a sad end to a satisfying partnership. However, we had become very good friends and had a common interest in sailing so, apart from ongoing work activities, we socialized quite a lot.

Al Greer and Dave Morgan were both taking every opportunity to gain experience as rapidly as possible. However, I sensed that Gordon Corps was a little uncomfortable that Al had 'been thrown in at the deep end' rather too quickly. Al was scheduled to do a series flight test on a HS 125 with me at Hawarden, near Chester, where BAe had their 125 production facility. Gordon had cleared it with the BAe test pilot for Al to carry out some additional flight tests to gain experience of the 125's handling characteristics. This was a useful opportunity for Al and he took full advantage of it. The aeroplane, straight off the production line, was fully fitted with the customer's interior layout. It was, therefore, somewhat heavier than a normal test aeroplane. The last items on Al's list of tests were the minimum control speeds. Even with the low fuel load towards the end of the flight, my calculations showed that we were in a situation where the aeroplane weight was such that the stalling speed at that weight was going to be quite close to the minimum control speeds. Normally, minimum control speed checks are made at much lower weights where the stalling speed is much lower and does not impinge on the test. At the higher weights, there is a potential danger. The minimum control speeds are assessed with one engine inoperative and the other engine at maximum take-off power, reducing speed slowly to the point where directional control can just be maintained with full rudder. When this is attempted close to the stalling speed, an inadvertent reduction of airspeed could result in the wing stalling. In this situation, with the very large thrust asymmetry, it is possible that the aeroplane could enter into a spin. As explained in the description of the NDN1 Firecracker flight-testing in Chapter 11, spinning is a high-risk situation and only likely to occur in single-engine light aeroplanes. The normal flight regime of a large transport aeroplane, even considering all the likely excesses, is such that spinning is a flight condition not required to be investigated. No-one knows what the consequences would be of an aeroplane like the HS 125 entering a spin.

Because we were low on fuel, Al was only going to assess the V_{MCA}, which is the take-off configuration case. Slowing the aeroplane down by gently pulling the nose up, applying increasing rudder deflection and sufficient bank angle to maintain a constant heading, Al was proceeding slowly. I was a little apprehensive as, I sensed, was the BAe test pilot. Al was concentrating. Then the stick-shaker (stall-warning) started to operate intermittently. Al was not known for giving a running commentary on his flying, so the other two of us were not party to his thoughts. I made a quick check of the scheduled stall warning speed for the weight we were at it and it was about 1 knot below the V_{MCA}. There were still a few knots of margin above the scheduled stall speed. Al progressed in silence, except for the short bursts of the stick-shaker. He was clearly determined to achieve the speed. The BAe test pilot was becoming twitchy; he had his left hand on the throttle lever of the live engine, initially to make sure it maintained full power but now, I was sure, he was ready to reduce power if necessary. His right hand was almost holding onto the control and once or twice, I thought he would call 'I have control' and take over.

In fact, Al managed to reach the V_{MCA} test point; he immediately pushed the nose down and the BAe test pilot reduced power on the live engine. The stick-shaker was quiet; at least two of us were breathing again. Al seemed very happy.

A couple of days later, back at the CAA when I was alone in the flight test engineers' office, Gordon came in, closed the door, and asked me how I thought Al had performed. I suspected he had already spoken with the BAe test pilot, but I could not be sure. I gave him a summary of the tests Al had accomplished and concluded with the circumstances of the final V_{MCA} test. I added:

'We all should have had the foresight to agree to break off the test a few knots above the V_{MCA}; it would have been no reflection on Al and a lot safer. I felt I should have been the one to make this case as soon as I had seen how close the stalling speed was going to be and I am disappointed with myself for not speaking up. However,' I added, 'to his credit, Al was concentrating very hard and he did nail the V_{MCA}, but he does seem to play his cards close to his chest and, watching him concentrate, it was impossible to judge how much realization he had of the big picture and its potential hazard.'

Gordon thanked me. He did not try to push me for any more detail and left without making any comment of his own; not like Gordon, I thought.

With not many months left, I was working hard to complete my backlog of flight test reports. Strictly, no-one could leave without fulfilling this task. In my case, the TriStar L-1011-500ACS saga, which had passed through numerous design changes, needed a lot of time to collate. Fortunately, I had made a start on the report after the first visit; this included many extracts from the Lockheed instrumentation records showing the pitch–up characteristics in detail. At the end of both the second and third visits, I collected a comprehensive sample of similar records from which I could choose some good examples to show the progress to a satisfactory outcome. In addition, these reports had to document all our flight test results, including those where the relevant requirements were fully complied with. This was necessary for two reasons. Firstly, if the CAA was invited to assess any significant design changes or a major development of the type, then we had a base-line of information for comparison. Secondly, much of this numerical information was useful in checking the 'flying characteristics' of the airline simulators against those of the real aeroplane; remember, the better the simulator, the more credit the airlines could take for the training in the simulator and fewer expensive training hours would be needed on the aeroplane. In the case of the L–1011–500ACS, all this data, amassed over the three phases of the flight testing, resulted in a very thick report; not easy reading for a non–flight-test person.

Al Greer had also been allocated the (now BAe) 748, so together we went up to Woodford to fly the latest version of the Series 2B. Despite their long allegiance to Smiths' autopilots, BAe Woodford had finally succumbed to the inevitable pressure from potential new operators and fitted the latest US Sperry autopilot/flight-director system. Additionally, they had decided to take advantage of the improved handling characteristics of the Series 2B and introduce a slight extension to the aft c.g. limit, thus improving the potential flexibility in the location of passengers and baggage.

We flew again with Charles Masefield. Geoff Stilgoe was the flight test engineer. Several days before he had been to Redhill for his interview, Keith Perrin had confirmed to me that he was the preferred candidate, but the letters had yet to be sent out and I agreed to say nothing. I managed a nod and a wink!

Al worked smoothly and confidently through the programme we had put together in the briefing, wielding his CAA spring-balance to good effect wherever necessary. One flight at the new aft c.g. limit was all that was necessary. The autopilot performance and the failure cases were no problem. The critical aft c.g. stability and control cases were back to being just as marginal as they always were on the Series 2A, but acceptable. Al also checked the V_{MCA} and V_{MCG}. For the latter, I had briefed him on the John Carrodus experience and that it was necessary to use some aileron. To his immense credit, on his first attempt, with copious use of aileron as well as rudder, Al managed to control the lateral deviation to a shade over the 30ft limit. His second attempt was comfortably within the deviation limit; he expressed himself satisfied and called it a day. BAe had pushed out the aft c.g. limit just as far as was reasonably possible. A good day for Al and my last Type Certification flight test; at least, that's what I thought at the time.

My penultimate flight was an airworthiness flight test on a British Caledonian Airways DC-10. After Gordon's departure, Al Greer would be taking over the responsibility for the McDonnell Douglas types. Gordon, I felt, was becoming a little obsessive about making sure Al could follow in his footsteps. Again, Gordon had agreed with the British Caledonian DC-10 chief pilot for Al to do a few additional handling tests that were not in the AFTS. This time Gordon came along with Al and they exchanged occupancy of the left-hand seat from time to time to share the flying. I had collected all the data necessary for the extra test cases and we had planned to fit them into convenient places within the AFTS testing. I don't know why Gordon did not let Al do all the flying and just monitor him from behind. Unlike the earlier BAe 125 test flight, there was nothing in this programme that should have presented any difficulty to an accomplished test pilot. Gordon had always prided himself on his ability to, almost seamlessly, lead from one test into the next, and I think he may have wanted to demonstrate this 'thinking ahead' process to Al. This certainly could reduce the necessary flight hours and, as I well remembered from my early flights with Gordon, it kept the flight test engineer on his toes. One of the aspects Gordon wanted to demonstrate was the stall handling. Like the Lockheed L-1011, the DC-10 also did not have a very pronounced nose-drop at the stall. If the pilot did not take stall-recovery action at that point, a notable wing-drop could result. Gordon wanted to 'calibrate' Al to be aware of where the 'CAA acceptable' nose-drop criteria lay and to show him the need to keep the sideslip under control to minimize wing-drops. With the benefit of Gordon's demonstration, Al carried out several stall tests in various configurations without difficulty. Everything else also went to plan.

For my final CAA test flight, Gordon had decided that I should finish in style, with a British Airways Concorde airworthiness flight test. I did not protest! I was completely ready for each test closely following the previous one, as the Concorde moved through the phases of flight from take-off, initial climb, transonic climb – with the engine after-burners on – and up to the cruise altitude of about 60,000ft at Mach 2.02. The over-speed warnings were always checked, which meant briefly reaching the maximum limit of Mach 2.12. Some further test points checked off and it was time to close the throttles for the descent back to lower altitude and subsonic flight, ensuring the automatic centre-of-gravity system was moving the fuel from the tail tank to bring it to the correct position for the subsonic flight. In this flight condition, we would carry out all the low-speed handling checks, including the stall warning and stick-wobbler operation. Finally, the visor and nose were lowered to give the pilot some reasonable view of the runway for landing, and there we were back on the ground at Heathrow.

Unusually, Gordon suggested that we call in at a local pub for a drink on the way home. Because we lived in different directions, we always used our own cars and normally departed on our separate ways. On this occasion, we had a very amiable couple of beers. Gordon was genuinely sorry that I had chosen to leave the department, but he thought working for Denis Murrin would be a good move; Denis was one of the few people in the CAA for whom he always had a good word. Gordon was himself a little sad about leaving but joining Airbus Industrie was the pinnacle of any test pilot's career (apart from, perhaps, becoming an astronaut). He was very happy to have handed the Airbus Industrie aeroplanes to Nick with John Denning continuing as flight test engineer, although I am not so sure he was relishing having to deal with the tenacious John Denning on the opposite side of the table. Keith was going to have a difficult time easing in two new FTEs at the same time. Gordon was happy with the choice of Geoff Stilgoe and I expressed my confidence that, with his experience in the civil aviation world, he would be a valuable asset and it would not take him long to be pulling his weight. Gordon still had several months more than me before he would depart and was still worried that it would leave Nick with a huge responsibility. He mentioned that wheels were in motion to recruit another test pilot. We reminisced about many interesting times and parted with something close to what has now become known as a 'man-hug'. It was an unexpected, pleasant surprise and I felt it was something that Gordon did not do regularly!

With some sadness, in the early spring of 1981, I left the Flight Department and moved all my personal office possessions up one floor, joining Andrew McClymont in an office adjacent to Denis Murrin, to start a new career direction with the title of design liaison surveyor. Since my early beginnings in flight testing at Avro in 1964, I had been a flight test observer/flight test engineer for seventeen years; not a bad innings. Furthermore, I had achieved my artificial goal in the CAA of one promotion every seven to eight years. It was much too soon to contemplate where I might go next; I had to make a success of this new job. I knew that Denis was due to retire in about five or six years' time, but Andrew would be expected to be the favoured successor and so I postponed any thoughts of future ambition until many years later.

A matter of days after I moved from the Flight Department, Geoff Stilgoe joined the CAA. Early in 1982 Gordon finally left for Airbus Industrie and Nick became the CAA chief test pilot. Shortly after this, a new test pilot, Mike Bell, joined the CAA flight test team; they were back to a reasonable complement of four test pilots (including Darrol) and four flight test engineers. However, only Nick Warner, Keith Perrin and John Denning had really significant experience of major evaluations. The Airbus variants were taking a lot of Nick and John's time. Al Greer and Dave Morgan had been working successfully together as a team on airworthiness flight tests and series checks. Nick had taken Geoff Stilgoe on similar flight tests. Keith concentrated on working with Mike Bell, as well as sharing his experience out among all the 'new boys' as best he could; the last thing Nick and Keith needed at that time was to have to commit a flight test team to a new foreign validation project when Nick and John Denning had a major Airbus commitment. It had to happen.

The German manufacturer Dornier had developed a new nineteen-seat, twin-turbo-propeller commuter aeroplane: the Dornier 228-100. In common with its competitors, it had a maximum weight limit of 5,700kg and therefore was designed to meet the FAR Part 23 and BCAR Section 'K' requirements. It had an advanced aerodynamic wing configuration and a moveable hinged tailplane with a conventional elevator. To trim the aeroplane for steady flight in any particular configuration, the angle of incidence of the tailplane was adjusted accordingly. Manoeuvring in pitch was accomplished with the elevator in the normal way. Aerodynamically, this arrangement resulted in less drag than with a conventional fixed tailplane/elevator combination as, in the steady trimmed condition, the elevator was always at the same angle as the tailplane.

In March 1982, Andrew McClymont took the evaluation team to Dornier. Al Greer and David Morgan would be the flight test team. Keith decided he needed to spare the time for him to go along as well, to oversee Al and Dave's review of the Dornier flight test results and their preparation of the CAA flight test programme. By then they had both been working at the CAA for about fifteen months. By comparison, it had been nearly two years before Nick and I were 'let loose on our own' for a major certification project; in our case, the Embraer 110 Bandeirante. That was after I had had prior first-hand experience of evaluating the type with John Carrodus.

The moveable trimming tailplane on the Dornier 228 was operated by an electric servo-motor controlled from the pilot's nose-up/nose-down trim switch. The Dornier engineers had provided an analytical assessment of the failure cases, indicating that a trim system runaway was very unlikely. Their worst-case likely failure scenario should only result in the trim system becoming inoperative. Consequently, the failure case testing for certification, carried out by Dornier, consisted of checking the manoeuvrability and ability to conduct a safe landing with the trim fixed at the most likely out-of-trim conditions. The CAA electrical/avionic engineer had reviewed the Dornier trim system failure analysis and, partly in consideration of his knowledge of similar systems on other types, he expressed

Dornier Do 228-100. (*ABPic, Paul Charles*)

reservations about the Dornier conclusions. Since the relevant report was in German, a firm decision was not reached by the CAA at the time.

Consequently, Al and Dave had an assessment of the handling and manoeuvrability with a nose-down out-of-trim in their forward c.g. flight programme. No maximum out-of-trim limit for the test had been specified by Dornier. Al started the test by trimming the aeroplane for level flight with appropriate power for an airspeed of 180 knots. He then operated the trim switch in the nose-down sense for about three seconds, while counteracting the out-of-trim by an increasing pull force on the control to maintain 180 knots. Apparently satisfied that this was not a problem, he continued trimming nose-down with the implication from his conversation with the Dornier pilot that he wanted to continue to full nose-down trim. Why the Dornier pilot did not question this and what Al's motivation for deviating from the test programme was will never be known. Additionally, the recorded instrumentation was continuously transmitted in real time by telemetry to the Dornier Flight Test Centre where it was being monitored. Although they were in communication with the pilot, the Flight Centre also did not pass on any comment about the deviation from the test programme. It was postulated afterwards that Al may have been trying to demonstrate that a full nose-down runaway was acceptable in order to avoid the need for the CAA electrical systems engineer requiring extra analysis.

Their problem began when Al found that, as the trim reached full travel, the elevator control forces were increasing excessively to about 150lb. He then tried to reduce the nose-down trim by moving the trim switch in the nose-up direction, but this had no effect. The aeroplane was pitching nose-down; the throttles were then closed but the speed increased rapidly. The two pilots were both pulling on the controls, generating a combined force estimated to have been over 300lb, but they could not recover the situation and the airspeed increased to over 300 knots (the maximum design speed was 250 knots) in a steep nose-down attitude. It started to break up at about 1,600ft altitude before impacting the ground. The detailed information above came from the Accident Report,[1] but I heard the salient details soon after the accident. If I had been the flight test engineer, I knew instinctively what I would have done: shouted at Al to roll the aeroplane upside down. Drastic, yes. The crew would be hanging in their straps but when inverted, the down elevator would be acting to raise the nose up, away from the ground, and reduce the airspeed. Without the load on the trim motor, the trim control would again be operable (trim servos stalling under high loads was not unheard of). The aeroplane could be rolled back the right way up and full control regained. A prolonged period of negative g, when upside down, could possibly upset the fuel supply to the engines but if either spluttered to a halt, there was probably enough altitude to accomplish an in-flight re-start. In the worst-case scenario, a controlled power-off landing on a field would have a very good chance of being survivable.

The reason why the trim servo-motor would not operate was because it had been disconnected by a torque-limiting clutch. This clutch was designed to limit the maximum electrical current to the motor and was set to operate at a torque equivalent to about 110lb control force. Al and Dave had not been made aware of this feature of the system design and, more surprisingly, it transpired, neither had the Dornier pilot.

Gordon Corps spent six years as an Airbus test pilot before it was time for him to retire from flying. Some time before, Airbus Industrie had identified Nick Warner as the ideal replacement; an offer he also was not going to turn down. Nick's arrival was almost coincidental with Gordon's retirement. Gordon was such a valuable 'aviation person' and not ready to put his feet up, so Airbus Industrie gave him a senior post in their Accident Investigation Department, a position he relished. Some years later, there was an Airbus fatal accident high in the Himalayas and Gordon was leading a team of Airbus engineers to the site to conduct an investigation. The only way to get close to the site was by helicopter. As the story goes, when the helicopter landed, Gordon, with his insatiable enthusiasm, was first out and bounded up the hill towards the crashed aeroplane. The altitude was very high and almost immediately, Gordon collapsed and succumbed to the fatal effects of altitude sickness. A very sad and unfortunate end to the career of one of the world's best test pilots. The huge number of well-known people from the aviation industry who attended his funeral was testament to this. I shed my own tear.

Later, in 1994 after Nick had been at Airbus Industrie for about six years, he lost his life together with all the crew in an Airbus A330 fatal accident at Toulouse. It is easy to say with hindsight, but this was an eminently avoidable accident. There were several tests on the flight programme, but the Airbus Industrie management had decided to combine the testing with an opportunity for familiarization of two pilots from the Italian airline Alitalia, and two Airbus employers were also on board as 'passengers'. Included in the scheduled tests was a simulated engine failure after take-off to assess a modification to the autopilot system. As I have made the point many times in previous chapters, flight testing of failures to automatic flight

Airbus A330. (*Avro Heritage Museum*)

control systems should be considered potentially 'high-risk'. This is particularly relevant for the complex autopilot and electronic fly-by-wire control system of the Airbus aeroplane types. It is highly likely that the decision to include the Alitalia pilots and other passengers on the flight gave those on board the sense that none of the scheduled tests would be critical.

Nick was the Airbus chief test pilot and on the day of this flight, he had already carried out an Airbus A321 demonstration flight, supervised a simulator session and attended two meetings including a press briefing. His co-pilot was not a test pilot but was an experienced Air-Inter training captain who had been working with the Airbus Industrie training organization, Aeroformation. The third member of the Airbus crew was Jean-Pierre Petit, a very experienced flight test engineer. There were the four other people on board, as mentioned above.

With Nick flying the aeroplane, they had completed a take-off and two go-arounds, each with a simulated engine failure. They then landed to make another take-off; this time with the modification to the autopilot control system. After the previous three successful tests, Nick was clearly not anticipating any problem and he gave control to the co-pilot. The co-pilot rotated the aeroplane more rapidly and to a higher nose-up attitude than normal. Nick was occupied with carrying out the test procedures of throttling back the port engine and de-selecting its hydraulic system (simulating the effect of the engine actually failing; this action had no significant influence on the hydraulic-powered flight control as there was sufficient redundancy in the hydraulic systems to deal with this failure) and engaging the autopilot. As the accident report said: 'He became temporarily out of the piloting loop.'[2] The autopilot was set to acquire an altitude of 2,000ft and it pitched the aeroplane more nose-up in an attempt to reach 2,000ft. It transpired that there was no pitch angle limit built into the autopilot computer in this situation. The airspeed rapidly decreased further to 100 knots (the V_{MCA} was 118 knots) and lateral control was lost. Nick realized what was happening, took control and reduced the thrust of the starboard engine to regain lateral control, but by then the bank angle was well over 90 degrees and the aeroplane was in a steep nose-down attitude. Although control was regained, it was too late to avoid ground impact.

Without going into further detail, this accident was caused by a combination of several factors, each avoidable. Knowing Nick as well as I did, the fact that he was ready to hand over control of the final test to the co-pilot demonstrated that he had been totally 'sucked into' the euphoria of complacency that had engulfed the essence of this test flight. He needed someone to remind him that there were potential dangers that may well require the awareness and skill of a competent test pilot. Another emotional funeral and an Airbus Industrie memorial service followed.

After I left the Flight Department, from time to time Darrol Stinton would seek me out and try to persuade me to accompany him for the spinning testing of various light aeroplanes. He was still finding the other flight test engineers rather unenthusiastic. I was usually busy with my new work so had always declined. However, some three years after I had closed my test flying log book, there he was again; the usual glint in his eye, and this time it was the latest version of the NDN Firecracker in which I had enjoyed flying so much with Darrol some five years earlier. I was tempted. As it happened, I had no serious work conflict at the time.

I checked with Denis Murrin who, being an ex-flight test engineer himself, said: 'Go for it, but make sure you book your man-hours and expenses to Flight Department and not my budget!'

The next day Darrol and I were off to Sandown airfield on the Isle of Wight, via the Hovercraft service from Southsea to Ryde. The aeroplane to be tested was the NDN T1, a turbine-engine version of the piston-engine type we had previously flown; lots more power and lighter in weight. After the normal trawl through the company test results, we put together our flight plan while being fed sandwiches and nibbles for lunch, and then off we went. At first, it was just like old times but, after the fifth or sixth spin, I was beginning to feel queasy. After two more spins, I had grabbed the sick-bag and filled it with my lunch. I was sitting immediately behind Darrol in the tandem seat and he could not see me. Stoically, and with the

NDN 1T Firecracker, Sandown, Isle of Wight, 1984. (*Author*)

judicious use of the intercom switch, I was able to carry on for a while without Darrol knowing of my problem. When I cautiously asked if we could have a short break, he realized straight away.

'I am terribly sorry. Do you want to go back?' he asked.

'No,' I replied. 'There are not many more spins to do and, if we take it slowly, I think I can continue OK. Besides, I have no lunch left to fill any more bags!'

Darrol offered to help by sharing the recording of some of the instrument readings. At a rather slower pace we completed the programme. I was unhappy that I had deprived Darrol of his opportunity to demonstrate his aerobatic prowess at the end of the flight. With more than a little embarrassment, I handed my sick-bag to the ground engineer after the flight; he was clearly quite used to it.

It had been a great surprise and disappointment to me that, with lack of use, my renowned constitution had deteriorated so noticeably. It is frequently said that one should never go back; good advice.

I had a new career that I was enjoying very much. There was no shortage of exciting and challenging responsibilities: the beginning of another of life's chapters.

Notes

1. Flugunfalluntersuchungsstelle, *Report on the accident to Dornier 228-100, D-IFNS near Aichach, FRG, March 26, 1982* (1983).
2. Commission of Investigation, *Accident on 30 June 1994 … Airbus A330 …* (1994).

A Brief History of the Evolution of Aircraft Safety Regulation in Europe

Throughout the period covered in this book, the British regulations and requirements for the safety of civil aircraft were administered by the Air Registration Board (ARB) until 1972 and then the UK Civil Aviation Authority (CAA). The Air Navigation Order (ANO) was, and still is, the primary legal UK legislation for the safety of aircraft and their operation. The detailed and prescriptive design requirements used to ensure the safety of aircraft were contained in the British Civil Airworthiness Requirements (BCAR), published initially by the ARB and latterly by the CAA. This was the situation throughout the period of this book.

However, it may be of interest to the reader to know the later developments in the regulation of aviation safety requirements. A summary is outlined below but a more comprehensive history is detailed in John Chaplin's paper.[1]

Sometime after the establishment of the European Union (EU), with a desire to standardize the requirements and procedures for aviation within the EU nations, a pan-European body was formed in the early 1980s called the Joint Aviation Authority (JAA). By 1987 it had expanded its responsibilities to include aircraft certification. Although it was never endowed with any legal status, all the national airworthiness authorities of the European nations gave it their support and endorsement. Under the auspices of the JAA, the member nations set about developing a set of standard requirements acceptable to the authorities of all the nations. These were known as the Joint Aviation Requirements (JAR). Almost all the European nations had been using the United States Federal Aviation Regulations (FAR) as the basis for their own requirements, the BCAR of the UK CAA being the most significant exception. Hence, as a basis for the development of the JAR for the design requirements, the relevant FAR was used as the starting-point. Needless to say, a great deal of negotiation was necessary for the UK CAA to convince the other nations of the validity of many of those individual BCAR that materially differed from the equivalent FAR. When initially published, the JAR contained several 'National Variants', these being individual requirements that some nations (mostly the UK) wished to retain. After several years of effort, which included a parallel requirement-harmonization programme between the JAA and the FAA, a consensus was reached, leaving only a very few national variants.

With a desire to build on and formalize this JAA arrangement, the EU nations agreed to set up a new body with legal status and the European Aviation Safety Agency (EASA) was formed in 2008. All the JAR were converted to European Aviation Safety 'Certification Specifications' (CS). Each nation has still retained its own national authority to administer some aspects of the CS in its own country, but many of the EASA processes, such as the Type Certification of new aircraft, were brought under the direct control and responsibility of the EASA.

NB – At the time of writing this book, the UK is negotiating to leave the EU. The consequences of this relating to the jurisdiction of EASA vs CAA is not known.

Note

1. Chaplin, *Safety Regulation: The First 100 Years* (2011).

A Summary of the Civil Aircraft Certification Processes

A civil aircraft may not fly unless it has been issued with a Certificate of Airworthiness (C of A) by the responsible Airworthiness Authority of the country in which it is registered (the State of Registry). Each such country has to be a signatory of the International Civil Aviation Organization (ICAO) and accepted by the ICAO as having the competence to maintain their own register of aircraft and issue such certificates. There are, of course, exceptions to this rule: for example, 'home-built' aircraft and some ex-military types which are permitted to be on the Civil Register. These will be approved to fly, by the Authority of the State of Registry, if they are deemed to meet an appropriate airworthiness standard and will be granted a lower-level certificate that may vary in different countries. In the UK, the CAA will issue a 'Permit to Fly', and in the USA, the FAA issues 'Experimental Certificates of Airworthiness' to such aircraft that are not eligible for a normal C of A. Such permits or certificates significantly restrict the operation and use of aircraft to which they are granted.

All aircraft of a type to which a full C of A may be granted must have been designed to meet all the appropriate airworthiness requirements or regulations as well as being constructed to defined manufacturing standards. These are specified by the Airworthiness Authority of the ICAO country in which the aircraft has been designed (the Authority of the State of Design) and where it has been manufactured (the Authority of the State of Manufacture); usually, but not necessarily, the same state. Such requirements or regulations must be accepted by the ICAO as meeting an appropriate minimum defined standard. In the UK, these were the British Civil Airworthiness Requirements (BCAR) until these were replaced by the Joint Aviation Requirements (JAR) developed by the European Joint Aviation Authority (JAA), and currently the Certification Specifications (CS) of the European Aviation Safety Agency (EASA). (See Appendix I.) In the USA, they are the Federal Aviation Regulations (FAR) of the Federal Aviation Administration (FAA). The ultimate confirmation that the Authority of the State of Design is satisfied that an aircraft type meets all the appropriate requirements is the issue of a Type Certificate by that authority.

In order to establish that an aircraft design meets these appropriate requirements, the aircraft must be constructed so that all necessary testing can be carried out to confirm this. Usually, more than one airframe will need to be manufactured to achieve all the necessary structural testing, system testing and flight testing. It is crucial that the airframe, including its installed systems and equipment, used for these tests and assessments must have been manufactured to a clearly-defined and documented design standard, otherwise the test results will not be valid. This standard is defined by the design company's Master Drawing for the aircraft type. It is crucial for the manufacturer to be able to establish, to the satisfaction of the Certificating Authority, that this is the case. The FAA uses the term 'conformity' to effectively describe this.

It is seldom a straightforward process. At this stage in the aircraft design definition, there is usually a steady stream of design alterations, each of which necessitates a change to the Master Drawing. There follows a continuous assessment of whether or not these changes will invalidate any testing already carried out on the current airframes; consequently, the tests may need to be repeated after the airframe has been modified. Concurrently, the manufacturer will usually be considering a list of options that will be available for potential customers. These could be alternative engines, passenger configurations, flight instrument systems, etc. or even differences in wing design or fuselage length. All these options must be added to the Master Drawing, and the airframes to be tested appropriately configured to encompass all such alternatives. At some stage in this initial design process it is necessary for the manufacturer to 'freeze' the Master Drawing, thus defining the envelope of the configuration(s) for which they require the Type Certificate to be granted by the Authority of the State of Design.

Once this (primary) Type Certificate has been issued, the Certificating Authority of any other ICAO country with potential customers for the type will have to issue their own Type Certificate which is a 'validation' of the Type Certificate of the State of Design. Where the Validating Authority has identical certification requirements to those of the Type Certificating Authority and has total confidence in the application of those requirements, they will issue their own validated Type Certificate as a 'rubber-stamp' of the primary Type Certificate. At the other extreme, when the Certificating Authority and the Validating Authority have significantly different requirements, some additional design investigation by the Validating Authority will usually be necessary. For instance, the UK BCAR were different to the FAR of the USA, although both met the required minimum standards of the ICAO. Even now, although the Certification Specifications of IASA (see Appendix I) are well harmonized with the FAR, there are still a number of differences and interpretations that warrant some additional investigation. Nevertheless, once the Validating Authority is satisfied that the aircraft complies with these differences, the principles of the ICAO require that it is the Certificating Authority who must make the statement that all the requirements, including the Validating Authority differences, have been shown to be met and the Validating Authority should accept this statement. With this statement, they are able to issue their validating Type Certificate.

For an authority to issue a Certificate of Airworthiness (C of A) to an aircraft on their register, the prerequisite is that they (the Authority of the State of Registry) must have granted a Type Certificate (primary or validation) for the type in question. Additionally, before a C of A can be issued, it must be established that the individual aircraft in question conforms to the aforementioned Master Drawing. Many validating authorities will not need to certificate every available design option or variant of the type and so the manufacturer will usually have a specific simplified Master Drawing for many of the individual Validating Authorities.

Once a Type Certificate is granted, there will always be a continuous process of improvements and other significant design changes that all need to be approved and incorporated into the Master Drawing. Keeping the Master Drawing up to date and ensuring that each aircraft conforms to this is an onerous but essential task.

Additionally, any country with its own competent design authority is entitled to approve modifications introduced by appropriately approved companies or engineers under its jurisdiction. For example, in the UK, companies such as British Airways hold a CAA (now EASA) approval that allows them to make design changes in certain defined areas to their Boeing or Airbus aeroplanes. These will generally be in the areas of passenger cabin configurations and equipment. Anything more complex will usually require design information and/or endorsement from the original manufacturer. Such 'third-party' modifications introduce complications when a second-hand aircraft, which incorporates locally-approved modifications, is imported. The importer has either to remove these modifications or find a way of showing them to be acceptable to their own national authority. Frequently, the act of removing a modification can leave the aircraft in a configuration not in conformity with the Master Drawing and hence will itself be a modification needing an approval.

Once a Type Certificate and subsequent Certificates of Airworthiness are issued, the process does not stand still. With even the best-designed aircraft, some faults and problems will arise from time to time during its service life. Any fault that may introduce a potential reduction in safety needs to be recorded and monitored.

The process of ensuring that an aircraft type retains the safety standards set at the time of certification is known as 'Continued Airworthiness'. Under the principles set out by the ICAO, each manufacturer responsible for the type design must establish, with every operator of the type, a system of recording, assessing and reporting such problems. All but the trivial items are routinely transmitted to the manufacturer's Airworthiness Authority, together with any actions proposed by the manufacturer.

If the problem has resulted in an accident or had a major impact on safety, this liaison with the Airworthiness Authority will be immediate. The manufacturer will have to propose a solution. This could be a design modification, a restriction to the operating limits (e.g. airspeeds, centre of gravity, etc.), increased maintenance checks or a combination of all three. These proposals will be considered by the authority and, if deemed to solve the problem, the authority will promulgate the solution as an 'Airworthiness Directive' (AD). Such ADs are mandatory, although they may include a time limit for compliance. If the problem is so serious that there is a high probability of a further such accident/incident occurring and no immediate rectification solution is available, then, in this extreme situation, the authority will have to ground all implicated aircraft. The easiest way to achieve this is to temporarily cancel all Certificates of Airworthiness. However, the manufacturer's authority has no direct jurisdiction over the issue of Certificates of Airworthiness by foreign authorities and, to protect their liability, their only recourse may be to temporarily withdraw the Type Certificate. Without the existence of a primary Type Certificate, the validated Type Certificates and the Certificates of Airworthiness issued by all authorities are invalid.

Most in-service problems fall in between this worst-case scenario and the 'no action necessary' category. The criteria generally used by the authority are: has the safety standard implicit in the applied certification regulations been eroded? Probabilities play a significant part in this decision. How frequently is the problem likely to occur and is this acceptable considering the severity of the event? When a modification is deemed to be necessary and an AD issued, the time allowed before compliance is implemented will generally be commensurate with probability considerations. Sometimes the manufacturer may take a tougher approach to the position of the authorities because they have to take account of 'product liability' and insurance issues. In many cases it is quite usual for the manufacturer to provide the modification free of charge.

Author's List of Flight-Tested Aircraft

AEROPLANE TYPE	Flight Hours	No. of Flights
Aerostar 600	4	2
Armstrong Whitworth Argosy C Mk 1	18	9
Avro (HS) 748 Series 1/2/2A/2B	425	258
Avro (HS) 780/748MF/Andover	142	94
Avro Shackleton Mk 2/Mk 3	18	10
Avro Viper-Shackleton	16	10
BAC 1-11 201/420/423/525/670	47	21
BAC Britannia 253	3	1
BAC/Aérospatiale Concorde	22	6
BAC VC10/Super VC10	22	6
Beech 60 Duke	1	1
Beech 76 Duchess	1	1
Boeing 707-320/321/336/349	11	5
Boeing 737-100	9	3
Boeing 747-136/236	10	3
Cessna 150F	2	2
Cessna 182P	1	1
Cessna 207	1	1
Cessna 310J	1	1
Cessna 336/337	4	3
Cessna 421B	2	1
Dassault Falcon 20E/F	19	8
Dassault Falcon 50	11	7
DH Tiger Moth	1	1
DHC-6 Twin Otter	6	4
Embraer Bandeirante EMB 110/E/K1/P/P2	36	23
Embraer Tucano	1	1
Embraer Xavante	1	1
Embraer Xingu EMB 121/B	41	23
Fokker F28-4000	20	9
Grumman American AA1B	1	1
Grumman Cougar GA7	1	1

AEROPLANE TYPE	Flight Hours	No. of Flights
Helio Courier H295	1	1
HP Herald 201/202/209/214	16	8
HP Victor K2 Tanker	73	32
HS 125-F400/600B/600F/700/700A/700B	32	17
HS 801 Nimrod MR1	411	164
HS Nimrod R	2	1
HS Trident 1/2/2E/3B	71	31
Learjet 25B	2	1
Lockheed TriStar L-1011-100/200/500/500ACS	218	59
Luscombe 8F	1	1
Lysander	0	1
McDonnell Douglas DC8-54F	6	1
McDonnell Douglas DC9-10	5	2
McDonnell Douglas DC10-30	25	8
Miles Student	1	1
Mooney M20E	1	2
NDN Firecracker 1/1T	6	4
Partenavia P68B	1	1
Piper PA-23-250 Aztec	23	16
Piper PA-31-325/350 Navajo	22	17
Piper PA-31T1 Cheyenne	9	6
Piper PA-32RT-300/T Cherokee	3	3
Piper PA-34-200T Seneca	3	1
Piper PA-39 Twin Comanche	1	1
Piper PA-42 Cheyenne III	12	7
Rallye 150T/ST	2	2
Rockwell 112/114	3	2
Rockwell 680F/685/690A	10	8
Rockwell 690C/695	12	8
Scottish Aviation Jetstream	1	1
Short SD 3-30	21	16
Vickers Viscount 708/722/814/838	15	8
Total aeroplane types = 64	**Total flight hours = 1,907**	**Total test flights = 949**

Note 1: 'Flight Hours' are rounded to the nearest hour.

Note 2: A 'Flight' is defined as commencing when the aircraft begins to taxi and terminating when it stops for the final time and the engine(s) are shut down. In some situations, a flight will be registered even if the aircraft never leaves the ground. For example, if a fault is encountered during taxiing which precludes getting airborne, or where the programme is limited to carrying out 'accelerate-stop' tests on the runway.

Glossary

Aeroplane	A heavier than air vehicle with fixed wings which generate lift when moving through the air.
Air Electronics Operator (AEO)	The flight crew member located in one of the rearward-facing seats of a 'V'-bomber (Valiant, Vulcan or Victor) who is responsible for managing those aeroplane systems not under direct control of the pilots.
AFTS (Airworthiness Flight Test Schedule)	A schedule of flight tests used by the CAA for each individual aircraft type when carrying out routine, in service air tests for C of A (see below) renewals or ongoing quality control.
Aileron	The moveable surface at the trailing edge of the outboard end of each wing that controls the rolling motion of the aeroplane.
Air brakes	See 'Speed brakes' below.
Aircraft	Any type of vehicle which can fly through the air. This includes aeroplanes, helicopters, balloons, airships and gliders.
Air Navigation Order (ANO)	The legal 'high-level' UK regulation, approved by Parliament, containing all the basic rules for flying in UK airspace. Other documents (such as CAP 360 and BCAR[1]) published by the CAA, provide the detailed interpretative material for aircraft manufacturers and operators to use.
Air Registration Board (ARB)	The body within the UK government responsible for the development, administration and implementation of aircraft safety requirements. It was formed in 1938 and its responsibility was transferred into the newly-formed Civil Aviation Authority (CAA) in 1972.
Airworthiness Directive (AD)	The means by which any National Airworthiness Authority can make mandatory a modification to a specific aircraft type that is deemed necessary for continuing safe operation.

Auxiliary Power Unit (APU)	An additional source providing electrical power; usually a turbine engine mounted in the tail of an aeroplane which drives a generator. Normally used to provide power on the ground but can also be started in flight for emergency situations.
AvP 970	The Ministry of Defence safety requirements having different sections for military aircraft of all types. (The MoD mission specifications are published in a separate document for each defined project.)
BAe (British Aerospace)	Formed in 1977 from the amalgamation of Hawker Siddeley Aviation, British Aircraft Corporation, Scottish Aviation and Hawker Siddeley Dynamics.
BCAR (British Civil Airworthiness Requirements)	The certification design requirements are in Section 'D' for large transport aeroplanes with maximum take-off weights of more than 12,500lb (5,700kg) and Section 'K' for all smaller aeroplanes. There are other sections for helicopters, gliders, airships, etc. Sections 'A' and 'B' contain procedural requirements for certification.
Bleed-air	High-pressure air extracted from a turbine engine, used for supplying the cabin pressurized air and other systems that are pneumatically-driven.
Civil Aviation Flying Unit (CAFU)	The body, being a part of the CAA and its predecessors, that was responsible for the calibration and approval of air navigation beacons and landing approach aids.
CAP 360	The document published by the CAA to give guidance to airlines and charter companies on how to comply with operational provisions of the ANO.
Centre of gravity (c.g.)	The position at which the aeroplane (or any object) would be balanced if it were suspended from this point.
Certificate of Airworthiness (C of A)	The certificate issued to an individual aircraft by the Airworthiness Authority of the State of Registry when satisfied that the aircraft conforms to the design standard specified in the Type Certificate (see below) and that all the prescribed maintenance has been accomplished.
Certification Specifications (CS)	The aircraft certification requirements published by the European Aviation Safety Agency (EASA); for example, CS-25 for large aeroplanes.

Ciência e Tecnologia Aeroespacial (CTA)	The Brazilian body responsible for the safety regulation and control of civil aircraft; equivalent to the UK CAA.
Control laws	All electronically-controlled systems, such as autopilots or stability augmentation systems, work by using all the relevant data to determine how best to move the flying control servo to achieve the required flight path. The programme that computes the signal to move the servo is known as the 'control law'.
Designated Engineering Representative (DER)	A status granted by the FAA to individual specialist engineers, either working for a company or freelance, which enables them to make findings of compliance with specific FAR as though they were directly employed by the FAA.
Direction générale de l'aviation civile (DGAC)	The French body responsible for the safety regulation and control of civil aircraft; equivalent to the UK CAA.
Direct Lift Control (DLC)	A system that uses hinged flaps (spoilers) on the upper surface of a wing to smooth out the wing lift in the presence of turbulence.
Elevator	The moveable surface fitted at the rear of the tailplane that controls the nose-up and down motion of the aeroplane.
European Aviation Safety Agency (EASA)	The body set up by the European Union nations in 2008 to develop and administer aviation safety regulations (Certification Specifications) within the European Union. (See also Appendix I.)
FAA (Federal Aviation Administration)	The body in the USA responsible for the safety regulation and control of civil aircraft; equivalent to the UK CAA.
FAR	US Federal Aviation Regulations; Part 25 for large transport aeroplanes and Part 23 for lighter aeroplanes.
Fin	The vertical aerodynamic surface mounted at the rear of the fuselage that provides directional stability.
Flight Management System (FMS)	A computer system into which all the information to define a planned route from take-off, climb, cruise and the descent to landing can be entered and which provides the guidance, usually through the autopilot, for the whole flight.

Flight Manual (often referred to as the AFM (Aircraft Flight Manual) or Airplane Flight Manual)	The document that provides the flight crew with the information necessary for safe operation of the aircraft. It includes limitations (weights, airspeeds, altitude, centre of gravity, etc.), emergency procedures, normal procedures and performance (take-off and landing distances and rates of climb).
Fly-by-wire	A design concept where the pilot controls the aircraft by electrical signals from the pilot's controls sent through a computer to the control servos without any mechanical connection.
g	A measure of force (usually in a direction normal to the horizontal axis of the aeroplane); one g is the force of gravity and zero g means weightless.
Glideslope	The angle measured above the horizontal down which an aircraft makes its approach to land (normally 3 degrees). Also refers to the radio beam transmitting in a horizontal plane at the approach angle. The signal from this is detected by a receiver on board the aircraft to provide indication of the degree of vertical deviation (up or down) from the glideslope.
GPS (Global Positioning System)	The system of satellites which transmit signals to a device (a GPS receiver) that uses the signals to compute and display the position of the device.
Gyroscope (gyro)	A rotating device that spins rapidly about an axis which in itself is free to alter its direction. The orientation of the axis is not affected by the tilting of the mounting so a gyroscope can be used to provide stability or maintain a reference in navigation or automatic pilot systems.
Hawker Siddeley Aviation (HSA)	Hawker Siddeley Aviation was a conglomerate of Hawker Aircraft, Armstrong Siddeley, A.V. Roe & Co., Armstrong Whitworth Aircraft, Gloster Aircraft Company, Folland Aircraft, de Havilland Aircraft Company and Blackburn Aircraft. In 1963, it became Hawker Siddeley Aviation and the constituent company names were dropped with their products rebranded as Hawker Siddeley (HS).
Hypoxia	The physiological effects of breathing air with too little oxygen; a condition experienced at high altitudes.

ICAO (International Civil Aviation Organization)	The international body, supported by all nations having an aviation industry, which formulates the internationally-accepted rules and principles for all aspects of aviation regulation.
Inertial Navigation System (INS)	A system of rapidly spinning gyroscopes (see above) which can detect small accelerations in all directions from which it calculates the distance and direction travelled from the starting location.
IMC (Instrument Meteorological Conditions)	Flight conditions where there is no external reference to give the pilot any information on the aircraft's attitude or location. The pilot must rely totally on the flight instruments.
IMN (Indicated Mach Number)	The Mach number indicated on the pilot's Mach-meter. This may differ slightly from the aeroplane's true Mach number because of errors arising from the location on the aeroplane of the altitude and airspeed pressure sensors from which the IMN is computed.
Incidence (angle of …)	The angle that the wing (or any aerodynamic surface) makes with the direction of the air in which it is moving.
Joint Aviation Authority (JAA)	The body set up in the early 1980s by the individual European Aviation Authorities to set aviation safety standards and administer those standards. It was superseded by the European Aviation Safety Agency (EASA).[1]
Joint Aviation Requirements (JAR)	The aviation safety requirements developed by the JAA for common application by all European National Authorities. These superseded the UK BCAR and were eventually replaced by the Certification Specifications of the EASA.
Knot	A speed of one nautical mile per hour. (1 knot is equal to approx. 1.15 miles per hour.)
Landing – Target Threshold Speed (V_{AT})	The scheduled landing target speed at the point 50ft above the runway surface. (See Chapter 4 for further information.)
Leading Edge Flap/ Slat	A moveable surface fitted on the leading edge of the wing that can be lowered to increase the lift of the wing at low speeds for take-off and landing. (See also 'Wing Flaps' below.)

Lift dumper	A moveable flap on the top of a wing which when raised significantly reduces the lift generated by the wing. Used to increase the rate of descent when necessary and on landing to ensure the aeroplane stays firmly on the ground for maximum wheel-brake effectiveness.
Localizer (transmitter)	The radio beam, transmitted in a vertical plane from the far end of a runway, from which a signal is detected by a receiver on board an aircraft. This signal provides an indication of the degree of lateral deviation (left or right) from the extended runway centreline.
Loop	An aerobatic manoeuvre which involves the aeroplane nose going up continuously until it becomes inverted and continues until it is back in horizontal flight.
Mach number	A number that expresses the speed of an aeroplane relative to the speed of sound. For example: Mach 1.0 is the speed of sound, Mach 0.8 is 80 per cent of the speed, and Mach 2.0 twice the speed of sound. The airspeed for Mach 1.0 at sea level is 761mph; it reduces with altitude to about 610mph at 36,000ft.
MAD (Magnetic Anomaly Detector)	A device for detecting small variations in the local magnetic field due to the presence of a metallic object. One of the sensors used for detecting submarines.
Minimum Control Speeds V_{MCA}, V_{MCL} & V_{MCG}	The lowest speed at which it is possible to maintain a constant heading using full application of the rudder control with one engine inoperative and the other(s) at maximum take-off power. V_{MCA} and V_{MCL} are the speeds in the take-off and landing configurations respectively and V_{MCG} is the speed on the ground prior to take-off. (See Chapter 4 for further details.)
Ministry of Defence (MoD)	The UK government department responsible for the procurement of military aircraft. This covers all phases from the issuing of a mission specification document, choice of manufacturer, setting of design criteria (e.g. AvP 970), overseeing the development and final acceptance.
Pitch attitude	The angle made by the longitudinal axis of an aircraft (nose-up or nose-down) to the horizontal.

Pitot-static	The pressure sensors on an aircraft that measure the dynamic and static pressure of the air through which the aircraft is flying. The pitot is a forward-facing tube into which the air flows to measure the dynamic pressure. The static pressure is measured from a hole (known as a port), usually located on the side of the fuselage, where there is no dynamic flow into the hole. The static pressure gives a direct reading of pressure altitude and the pitot pressure combined with the static pressure computes the airspeed.
Ram Air Turbine (RAT)	A device that can be lowered from an aircraft in flight; it has a small propeller that drives an electric generator and/or a hydraulic pump. It is used in serious emergency situations such as loss of electrical and/or hydraulic power.
Recovery Speed Brake (RSB)	A system that deploys the speed brake panels on the upper wing surface to reduce the wing lift and increase the aerodynamic drag.
Rijksluchtvaartdienst (RLD)	The Aviation Authority of the Netherlands responsible for the safety regulation and control of civil aircraft; equivalent to the UK CAA.
Roll attitude	The angle made by the lateral axis of an aircraft (left-wing-down or right-wing-down) with the horizon.
Rudder	The moveable surface at the rear of the fin that controls the direction of the nose relative to the airflow (sideslip).
Runway Visual Range (RVR)	The range over which the pilot of an aircraft on the centreline of a runway can see the runway surface markings or the lights delineating the runway or identifying its centreline.
Satellite Navigation System (Satnav)	See 'GPS' above.
Servo	An electric or hydraulic-powered motor that moves the aircraft control surfaces.
shp	Shaft horse-power of a propeller engine; the horse-power measured at the propeller shaft.
Sideslip	The horizontal angle (left or right) which the fore and aft axis of the fuselage make with the direction of air in which it is moving.

Sinusoidal	An oscillating motion that follows a regular trajectory in the shape of a 'sine-wave'. Radio waves and light waves follow a 'sine-wave' motion.
Slide rule	Before the advent of hand-held calculators, this was a device routinely used to do quite complex mathematical calculations.
Speed brakes (also known as air brakes)	Aerodynamic panels that can be extended (usually from the upper wing surface but sometimes from the rear fuselage) which are designed to create a lot of drag to slow the aeroplane down.
Special Federal Air Regulation (SFAR)	A means by which the FAA can mandate a requirement for some specific training or operational procedure. Similar in principle to an Airworthiness Directive (AD) which is used for mandating design changes.
Spinning	Aeroplanes can enter a spin from low speed; typically, close to the stalling speed, with a large input of rudder. This can happen inadvertently if the pilot is over-ambitiously trying to control sideslip or making a tight turn at very low speeds. In the resulting spin, the aeroplane takes up a stable nose-down attitude and rotates quite rapidly about the vertical axis; being effectively stalled, the speed does not increase. Recovery is usually achieved by applying full opposite rudder but sometimes elevator and/or aileron may also be necessary. Spinning is almost uniquely a problem associated with light aeroplanes.
Stability Augmentation System (SAS)	An electronic system which, through a servo-motor, adds inputs to the control circuit in order to enhance the natural stability of the aeroplane.
Stall	The condition reached by a wing (or any aerodynamic surface) when its angle of incidence to the local airflow is so great that it can no longer generate sufficient aerodynamic lift to support the aeroplane.
Static (port or pressure)	See pitot-static above.
Stick-pusher	An electro/mechanical device that pushes the pilot's control column forward to lower the nose and increase airspeed to prevent the aeroplane reaching a dangerous stall condition.

Stick-shaker	An electrically-operated device which rapidly and energetically shakes the pilot's control column to warn the pilot of an imminent stall. It simulates the aerodynamic buffeting of the aeroplane that may occur prior to reaching a stalled condition.
Stick-wobbler	A similar device to a stick-pusher installed on Concorde that effectively limits the minimum airspeed to a safe value (known as V_{MIN}).
Tab (geared or spring) (see also trim tab below)	A small control surface on the trailing edge of an elevator, aileron or rudder which is used to vary the amount of control force needed to be applied by the pilot to achieve a particular control deflection.
Tailplane	The horizontal surface (small wing-like structure) mounted behind the wing. (See also 'elevator'.)
Take-off Safety Speed (V_2)	The minimum speed, scheduled in the Aeroplane Flight Manual, which the aeroplane should achieve at a height of 35ft after take-off. (See Chapter 4 for more information.)
Threshold (of runway)	The location on a runway that defines the beginning of the designated landing zone. This is usually about 1,000ft before the point at which the approach path of an aeroplane (normally 3 degrees from horizontal) intersects with the runway. This is to allow a margin of safety for inadvertent under-shooting of this point.
Trim tab	The aerodynamic surface mounted at the rear of an elevator, aileron or rudder which when moved by a trim-wheel or switch operated by the pilot exerts an aerodynamic force on the control to remove any residual control force applied by the pilot.
Type Certificate	The certificate granted by an Airworthiness Authority to an aircraft manufacturer signifying that the aircraft type listed on the certificate meets all the airworthiness requirements specified by the authority. The certificate must define, usually by reference to a Master Drawing, the build standard of the aircraft for which the compliance has been shown. This includes all variations of the type and any modifications within the envelope of the type design. (See also Appendix II.)

Vortex generator	Thin metal 'blades' fitted onto the surface of a wing, tailplane, rudder or control surface. They are set at a slight angle to the local airflow and have the effect of re-energizing the air close to the surface to improve the aerodynamic effectiveness. A simple idea used on many aircraft (as well as other vehicles), it was successfully patented by John D. Lee in 1953, giving him a royalty income for each individual one fitted.
Wing flap	The lowerable and extendable sections at the trailing edge of a wing which are used to increase the lift of the wing at low speeds for take-off and landing. (See also leading-edge flaps.)
Yaw	The angular rotation (nose-left or nose-right) that an aircraft makes with the vertical axis.

Note

1. Chaplin, *Safety Regulation: The First 100 Years* (2011).

Bibliography

Blackman, A.L., *Test Pilot: My Extraordinary Life in Flight* (London: Grub Street, 2009)

Blackman, A.L., *Victor Boys: True Stories from Forty Memorable Years of the Last V Bomber* (London: Grub Street, 2012)

Chaplin, J.C., *Safety Regulation: The First 100 Years* (Paper Number 2011/3 of the *Journal of Aeronautical History*, The Royal Aeronautical Society, 2011)

Commission of Investigation, *Accident on 30 June 1994 … Airbus A330 No 42 of Airbus Industrie* (Bielefeld: University of Bielefeld, 1994)

Fildes, D.W., *The Avro Type 698 Vulcan: The Secrets Behind its Design and Development* (Barnsley: Pen and Sword, 2012)

Flugunfalluntersuchungsstelle, *Report on the accident to Dornier 228-100, D-IFNS near Aichach, FRG, March 26, 1982* (Braunschweig: Flugunfalluntersuchungsstelle, 1983)

McDaniel, J.D., *Tales of the Cheshire Planes* (Peterborough: GSM Enterprises, 1998)

White, R., *Vulcan 607: The Epic Story of the Most Remarkable British Air Attack Since WWII* (London: Transworld Publishers, 2006; Corgi, 2007)

Index

Aberdeen, 134
Aerolineas Argentinas, 183
Air Anglia, 209, 212
Airbus Industries, 186, 234, 247–8,
  250–2
  A300, 186
  A321, 252
  A330, 251–2
Air France, 157, 159, 171
Air Malawi, 183
Air Registration Board, 118
Alitalia, 251–2
Amsterdam, 130, 150, 209
Anglesey, N. Wales, 116
Arctic Circle, 134
Armstrong Whitworth, 61, 69
  Argosy, 72–3
Ashford, Ronald, 234, 236
Asmara, Ethiopia, 82–6
Austin 7, 37
Austin Mini/Cooper/'S', 67, 77–8, 105,
  118, 124
Avro (A.V Roe & Co), 1, 28–9, 38, 41,
  44, 60–1, 90
  707, 44
  730, 42–3
  748, 4, 28, 31, 44–5, 51, 58, 61–2, 64,
    71–4, 81, 89–90, 92, 94–104, 109,
    114, 122, 125, 155, 172, 183, 190,
    200, 206, 212, 214, 223, 245
  780 (748MF/ Andover), 44, 58–64,
    68, 72–4, 79, 89, 110–13
  Shackleton, 31, 68–71, 79, 93–4
  Vulcan, 2–4, 28, 31, 41–2, 44, 58–62,
    68, 114, 143
Avro (Woodford) Motor Club, 76

BAC/Aerospacial Concorde, 42, 141–3,
  147, 157–9, 171–3, 187, 219, 246
Bailey, Tony, 108, 116
Bedford (MoD/RAEE), 71, 111
Beech, 144
Bell, Mike, 247–8
Benoy, Mike, 192, 215, 235
Bentley, Derek, 60, 67, 76–7, 95, 104,
  110
Bethany, Oklahoma, 215–16
Bethpage, New York State, 190
Blackburn Aircraft Company, 28
  Buccaneer, 28, 114
Blackman, Tony, 1, 9–10, 73, 81, 87–8,
  92, 98, 105, 214
Blum, Phil, 177–8
Bodo, Norway, 134
Boeing;
  707, 125, 134, 140, 155, 169, 209
  727, 41, 150, 224
  747, 209
Bollington, Cheshire, 118
Boor, Reg, 113
Bordeaux, 213, 225
Boscombe Down, 65, 139, 186–7,
  234
Bournemouth, Hurn, 184–5
Brabazon House, 118, 120
Bristol, 29, 36–7, 122
  University, 29, 32, 34–5, 43
British Aircraft Corporation,
  1–11, 125, 138–40, 148, 150, 155
  Harrier, 105–106, 139
  Hawk, 105–106
  Jet Provost, 69
  VC10, 114, 125, 139, 150, 183

British Airways, 123, 140, 171, 174, 176–8, 183–4, 190, 207, 220–1, 231, 239

British Caledonian Airways, 169, 246

British West Indian Airways, 236, 240

Brough, E. Yorkshire, 28

Burns, Don, 119, 131, 144, 146, 154–5, 171, 180, 182, 186, 207, 220

Cabral, 160, 162, 169–70, 189, 198, 202

Cairo, 82, 87

Carrodus, John, 87–9, 93, 105, 111, 119, 123–5, 139, 144, 146–8, 152, 154–8, 160, 162, 168–9, 180–2, 207, 220, 240, 245, 248

Casablanca, Morocco, 171–2

Casbard, Bob, 158

Cathay Pacific Airways, 154

Cessna, 127
    150, 127
    310, 129
    336/337, 128–9

Chadderton, 28–9, 43

Chaplin, John, 238, 255

Cherry-Downes, Steven, 82, 84

Chester (de Havilland/HSA), 79, 98

Chevrolet Camaro, 205

Chicago, 177

China, 45, 220

Collins, 91–2

Concorde,
    See BAC/Aerospacial Concorde

Corps, Gordon, 74, 119, 121–3, 125–7, 139, 141, 150–2, 155, 171–3, 176–8, 183, 186, 190–1, 193–7, 201, 205, 212, 214–15, 219, 223, 225, 234, 236, 243–4, 246–7, 250–1

Cranfield, 169–70

Cretney, Dave, 183, 221–2

Creykes Sidings, 14

Cruise, John, 114

CSE Aviation, Kidlington, 169, 187

Cummings, Dave, 119, 127, 171, 186, 234

Cunningham, John, 79

Dakar, W. Africa, 158–9

Dan-Air, 183, 190, 200

Dane's Moss, 14, 16

Dassault, 148, 155, 224–5
    Falcon 20, 201, 213
    Falcon 50, 224–6

Davies, Dave, 91, 115, 119, 121, 125, 131–2, 134, 137, 141, 146, 155–6, 181, 183, 186, 214, 235

Death Valley, 239–40

Decca-Navigator, 99–101

de Havilland, 61, 79
    Comet, 69, 79, 94
    Dove, 82
    Tiger Moth, 135–6
    Venom, 34
    Canada DHC-6 Twin Otter, 125–7

Delta Airlines, 177, 181

de Mercator, Roger, 154

Denning, John, 186, 234, 248

Digital, 95, 101–102

Dixon-Stubbs, Bob, 82–3

Dornier, 248–9
    228, 248–9

Dublin, 130, 183

EASAMS, 109

Edwards Air Force Base, California, 205–206, 211, 222

Eifflander, Gig, 20, 22

Elliott, Carol, 214

Else, Bill, 69–70, 72, 94, 111, 113, 168

Elstree airfield, 173, 175

Embraer, 155, 157, 159, 161, 165, 187, 193, 200, 204, 240
    Bandeirante (110), 155, 158–9, 162–4, 169–70, 187, 240–1
    Ipanema, 167–8
    Tucano, 241–2
    Xavante, 163–4
    Xingu (121), 197–203, 240–2

Empire Test Pilot's School (ETPS), 139, 214, 234

English Electric Lightning, 114

Falklands War, 114
Farnborough Air Show, 187
Fisher, Harry, 5, 82, 84, 114
Fokker, 4, 148, 212
    F27, 4
    F28-4000, 148–52, 209, 212–13
Ford Capri 3000, 124, 133
Ford Cortina, 105, 124
Fox, Gordon, 82, 87
Franklin, Eric, 72
Furnace Creek Inn, 239

Garland, Martin, 79, 95
Garrett TFE731, 152–3, 224
Garrett TPE331, 215
Gettysburg, Pennsylvania, 195–6
Gibbons, Dave, 111
Glaser, Dave, 138
Glos Air, 184–5
Gollings, Dave, 184–5
Goodfellow, Mike, 154
Goodwood airfield, 138
Goole, Yorkshire, 13–14
    Goole Grange, 14
Greenland, 176
Greer, Al, 234, 243–6, 248–50
Grieve, Stuart, 92
Grimsby, 100
Ground Proximity Warning (GPWS),
    125, 147
Grumman Gulfstream, 190, 196
Guido, Pessoti, 159, 165
Gulf Air, 184

Hadwen, Rodney, 38–9, 42–3
Hall, Laurie, 119
Handley-Page, 4, 7
    Herald, 4
    Victor, 1–8, 10–11, 109–10, 113–15,
        117
Harbin Y-12, 220
Harper, Peter, 187
Harrison, Jimmy, 62–3, 72, 79, 81, 92
Hartley, Ted, 5, 62, 95
Hatfield, 93, 106, 152, 154, 161, 197
Hawker Siddeley Aviation, 4, 61, 74, 79,
    104–105, 110, 117, 148

125, 69, 123, 152–4, 200, 223, 240,
    243
146, 106–107
Nimrod, 69, 71, 80–1, 94–8, 100–104,
    109, 155
Trident, 41, 90, 150, 183, 190, 197,
    224
Hawker, 61
    Typhoon, 69
Heathrow, 93, 246
Helio Courier H-295, 136–7
Hillman Imp, 108
Hoover, Bob, 217
Horsham, Sussex, 118
Horsley, Bill, 71, 81, 86, 108–109, 113,
    115, 118–19, 131, 186, 198, 215, 220,
    238–41
Hyodo, 198

ICI (Imperial Chemical Industries), 76
Iguazu Falls, 165–7
Indian Air Force, 104
Iroquois catamaran, 117
Isle of Wight, 115, 253
Istres, S of France, 201, 225

Jacobson, Carl, 207, 211, 237

Keegan, Kevin, 174–5
Kennedy, Austin, 219
Kingston, Surrey, 105, 113
Kirton, Ruby, 118–19
Kitchen, Sally, 41

Lakeland, Florida, 222, 235
Laker Airways, 204
Lakin, Terry, 183, 221–2
Lancaster, California, 179, 205, 221–2
Lawton, Bill, 193–4
Le Bourget (Paris), 173–4
Leckman, Paul, 216–17
Ledwidge, Flight Lieutenant, 111, 113
Learjet, 183
Lockhaven, Pennsylvania, 190–1, 194
Lockheed, 193, 205, 207, 209, 226–33,
    239
    Hercules, 114

TriStar L1011, 155, 178, 180–4, 190, 204, 207–208, 211, 220, 224, 226–33, 236–9, 244-5
Long Beach, California, 176–8, 204–205
Los Angeles, 176–8, 181, 204–205
Luscombe 8F, 127–8
Lycoming 0-540, 201
Lysander, 135

Macclesfield, Cheshire, 13–15, 20, 108, 115
   Grammar School, 25
   Model Aircraft Club, 22
Malta, 87
Manchester, 76–7, 98
Masefield, Charles, 5, 114, 214, 245
Massawa, Ethiopia (now Eritrea), 82, 84
McClymont, Andrew, 62, 71, 81–2, 86, 119, 133, 157, 185–6, 192, 219–20, 235, 247–8
McDaniel, John, 63, 87, 104
McDonnel Douglas, 173, 177, 193, 204, 246
   DC-8, 173–4, 209
   DC-9, 134
   DC-10, 175–6, 204, 209, 237, 246
Meccano, 18
Merlin Rocket, 108, 116
Miami, 181–2, 224
Mitsubishi MU-2, 145, 147
Mig 21, 104
Millar, Flight Lieutenant, 111
Ministry of Defence (MoD), 2–3, 44–5, 60–1, 65, 72, 74, 79, 97, 101, 114
Mooney M20, 127
Moor, Don, 208
Morgan, Dave, 234, 238–40, 243, 248–50
Morgan, Peter, 98, 109
Morris 8, 37
Morris Cowley, 24
Murrin, Denis, 157–61, 164–5, 168, 187–8, 190, 192–3, 195–8, 214–15, 219–20, 222, 226, 235–36, 241, 247, 253

NDN -1 Firecracker, 200–201, 209–10, 243, 253

New York, JFK, 190–1, 196
Nice, 87
Nicosia, 82

Ogilvie, Squadron Leader, 79
Olenski, Zbigniew, 42
Oslo, 130, 134

Palmdale, California, 155, 178, 182–3, 204–205, 207–208, 211, 220, 226, 236–8
Partenavia P68, 129
Perrin, Keith, 74, 87–9, 114–15, 119, 139, 141, 144, 147, 155, 171, 173, 175, 185–6, 204, 234–6, 238, 245, 247
Perry, Brian, 157, 167
Petit, Jean-Pierre, 252
Piper, 127, 144, 155, 190, 192–3
   PA-23 (Aztec), 129, 134, 173, 190–1
   PA-25 (Pawnee), 167
   PA-31 (Cheyenne), 190–2, 194, 198
   PA-32 (Cherokee), 200
   PA-34 (Seneca), 127
   PA-42 (Cheyenne III), 222, 224, 235
Plymouth (RAF Mountbatten), 74
Pogson, Bob, 5, 80, 95
Poole, Dorset, 116–17
Pratt & Whitney
   PT6A, 158, 169, 190, 222
Progress Aero Works (PAW), 20

Radlet, 3
Raisal, Pierre, 201
RAF, 28, 60, 79, 97, 103, 110, 114, 243
Redhill, Surrey, 115, 118, 120, 146, 190, 231, 245
Reed, Ken, 186
Richards, Peter, 157, 187, 198, 241
Ringway (Manchester Airport), 18
Rio de Janeiro, 157–9, 187, 203
Roberts, Geoff, 90
Robin 200/100, 138
Rockwell, 155, 185–6, 215
   112, 129
   685, 121, 123, 139–40
   690, 184–5, 215

695, 215–16
    Shrike Commander, 217
Rocky Mountains, California, 228, 230
Rolls-Royce, 184, 237
    Olympus, 143
    Spey, 149
    RB211, 237
    Viper, 69, 105, 152
Rome, Ciampino, 82
Russia (Soviet Union), 45, 130, 132
Rust, Bert, 192, 198, 215, 224, 241
Rye, John, 147–8

Safety Assessment of Systems, 58,
    217–18, 228, 231, 233
Salford Technical College, 31
Sao Jose dos Campos, 157, 159, 162,
    167–8, 187–9, 197–8, 242
Scard, Doug, 104
Scottish Western Isles, 134
Semark, Flight Lieutenant, 111
Sergio, 202, 242
Shoreham airfield, 138
Shorts (Belfast), 133–4, 220
    SD3-30, 133
Simulators, 129–30, 209, 212, 214, 245
Singer Gazelle, 40, 77
Skillen, Graham, 220, 236
Skybolt, 58–60
Smith, Ivor, 15–17, 20, 22–3
Smith, John (Dan Air), 190, 200
Smith, Leslie, 14
Smith, Mike, 86–7, 119, 186
Smith, Stan, 13–14
Smith, Will, 180
Smith's industries, 7, 74, 90, 92, 212–14
Socata Rallye 150, 129
Southampton, 100
Sperry, 245
St Mawgan, RAF, 98
Stavanger, 134
Staverton, Gloucestershire, 121
Stilgoe, Geoff, 90, 238, 245, 247–8
Stinton, Darrol, 119, 126–7, 134–6,
    138, 148, 162, 186, 200, 209–10, 215,
    253–4
Sunbeam Talbot 10, 37, 39–40

Talbot (Hillman) Avenger, 132
Taylor, Mike, 82, 84
Themen, Henk, 150
Thruxton, 128
Tomlinson, Geoff, 32–4, 36, 39–40
Toulouse, France, 234, 251
Trondheim, Norway, 134
Trubshaw, Brian, 172–3
Turner, Colin, 158, 167
Turner, Mike, 62–3, 72, 79–82, 84

Vickers;
    Valiant, 2–3
    Viscount, 138
Vincent, Alan, 109

Warner, Nick, 139–41, 155–6, 169–70,
    178–82, 187, 189, 198, 204–206,
    214–17, 220, 226–32, 234, 239–43,
    247–52
Warren, Denis, 147–48
Washington (FAA Head Office), 224
Waterpark, Lord, 187
Warton (Bae), 139
Weaver, Bill, 180, 227
Webster, John, 183, 221
Wells, John, 180
Wild Goose, 178, 205, 221
Wills, Vin, 157, 192, 215, 241
Wilson, Bob, 115
Woburn Abbey, 113
Wood, Harry, 76
Wood, Ken, 44, 63–4, 79
Woodford, Cheshire, 1, 10–11, 28–31,
    58, 69, 74, 79, 98, 103–104, 106, 113,
    115, 151, 212, 214, 245
Wyrick, Sam, 180, 182, 205, 226–7,
    229, 232
Wyton, RAF, 5

Yak 40, 130–2
Yuma, Arizona, 204